FINDING BALANCE 2023

BENCHMARKING PERFORMANCE AND BUILDING CLIMATE RESILIENCE IN PACIFIC STATE-OWNED ENTERPRISES

MARCH 2023

© 2023 Asian Development Bank
6 ADB Avenue, Mandaluyong City, 1550 Metro Manila, Philippines
Tel +63 2 8632 4444; Fax +63 2 8636 2444
www.adb.org

Some rights reserved. Published in March 2023.

ISBN 978-92-9270-058-4 (print), 978-92-9270-059-1 (electronic), 978-92-9270-060-7 (ebook)
Publication Stock No. TCS230063-2
DOI: http://dx.doi.org/10.22617/TCS230063-2

The views expressed in this publication are those of the authors and do not necessarily reflect the views and policies of the Asian Development Bank (ADB) or its Board of Governors or the governments they represent.

ADB does not guarantee the accuracy of the data included in this publication and accepts no responsibility for any consequence of their use. The mention of specific companies or products of manufacturers does not imply that they are endorsed or recommended by ADB in preference to others of a similar nature that are not mentioned.

By making any designation of or reference to a particular territory or geographic area, or by using the term "country" in this document, ADB does not intend to make any judgments as to the legal or other status of any territory or area.

Please contact pubsmarketing@adb.org if you have questions or comments with respect to content, or if you wish to obtain copyright permission for your intended use that does not fall within these terms, or for permission to use the ADB logo.

Corrigenda to ADB publications may be found at http://www.adb.org/publications/corrigenda.

Notes:
In this publication, "$" refers to United States dollars.

Cover design by Oblivious Design & Print.

Contents

Tables, Figures, and Boxes

Tables

Figures

Boxes

Foreword

Pacific island countries recognize the importance of trade, investment, and private sector development to achieve inclusive economic growth. Central to the region's development efforts is the commercialization of state-owned enterprises (SOEs), which remain a dominant feature of Pacific economies and provide core infrastructure services, such as airports and ports, power, water supply, and sanitation.

Reforms to date have improved SOE performance in the Pacific by strengthening the legal, regulatory, and governance frameworks under which they operate. Reforms have also encouraged SOE partnerships with the Pacific private sector, stoking investment and innovation. However, as governments in the Pacific work to improve the financial sustainability of their SOEs, they also must incorporate strategies that address emerging risks, such as climate change. Climate scientists expect that the Pacific will confront more intense and frequent natural hazards, such as cyclones, floods, and coastal inundation, as global temperatures increase. SOEs will thus be obligated to mitigate the financial risks associated with climate change, and to protect their assets—many of which are critical for the functioning of society—from climate-induced physical damage.

This is the seventh benchmarking study in the Finding Balance series, first published in 2009, which aims to track the progress and impact of SOE commercialization in participating Pacific countries. Commercialization is measured by the financial performance of the SOEs and their supporting legal, regulatory, and governance environment; the study does not benchmark technical performance. In addition to its survey of SOE financial performance, this edition has a special focus on the risks posed by climate change, and how governments and their state-owned utilities are acting to mitigate its effects and build resilience.

Finding Balance 2023 posits that Pacific state-owned utilities that are more commercialized may be more inclined to respond to incentives to decarbonize and invest in protecting their assets. However, to optimize these choices, SOEs must operate in a policy environment conducive to commercialization. The study also shows that government ownership and management of SOEs are not essential for public service delivery. As the experience of Fiji, the Marshall Islands, Papua New Guinea, Samoa, and Vanuatu demonstrate, partnerships with the private sector can expand the capacity of governments to deliver public services efficiently. A key theme of this study is finding the balance between the roles of the public and private sectors.

The surveyed countries—Fiji, Kiribati, the Marshall Islands, Palau, Papua New Guinea, Samoa, Solomon Islands, Tonga, and Vanuatu—were chosen for their willingness to identify and address the core issues within their SOE sectors, including in relation to climate change impacts. I thank the governments of all participating countries for their input and support.

I also wish to thank the authors of the report, Laure Darcy, Arjuna Dibley, Brent Fisse, and Christopher Russell, analysts David Ling, Alma Pekmezovic, and Minh Vu, and editor Angelo Risso for their efforts, as well as the governments of Australia and New Zealand, which provide cofinancing under the Pacific Private Sector Development Initiative (PSDI).

I am confident *Finding Balance 2023* will provide thought-provoking insights and stimulate useful discussions on the role that SOEs can play in building resilience to climate change in the Pacific.

Leah Gutierrez
Director General, Pacific Department
Asian Development Bank

Abbreviations

ADB	Asian Development Bank
AKA	Airports Kiribati Authority
AMI	Air Marshall Islands
AV	Airports Vanuatu
AVOL	Air Vanuatu (Operations) Limited
BSCC	Belau Submarine Cable Corporation
CEMA	Commodities Export Marketing Authority
COVID-19	coronavirus disease
CSA	commercial statutory authority
CSO	community service obligation
DBK	Development Bank of Kiribati
DBS	Development Bank of Samoa
EFL	Energy Fiji Limited
EPC	Electric Power Corporation
FCCC	Fiji Competition and Consumer Commission
FM Act	Financial Management Act 2004 (Fiji)
FNPF	Fiji National Provident Fund
FPCL	Fiji Ports Corporation Limited
FY	fiscal year
GBMEU	Government Business Monitoring and Evaluation Unit
GBT	General Business Trust
GCC	government commercial company
GDP	gross domestic product
GHG	greenhouse gas
GWh	gigawatt-hour
ICSI	Investment Corporation Solomon Islands
IPP	independent power producer
KAJUR	Kwajalein Atoll Joint Utilities Resources
KCH	Kumul Consolidated Holdings
KHC	Kiribati Housing Corporation
KNSL	Kiribati National Shipping Limited
KOC	Kiribati Oil Company
KPA	Kiribati Ports Authority
MEC	Marshalls Energy Company
MPE	Ministry of Public Enterprises
MRI	Majuro Resort Inc.
MVIL	Motor Vehicles Insurance Limited
MW	megawatt
MWh	megawatt-hour
NBV	National Bank of Vanuatu

NCO	noncommercial obligation
NDBP	National Development Bank of Palau
NDC	Nationally Determined Contributions
NEC	National Executive Council
NHC	National Housing Corporation
NPAT	net profit after tax
PacDMC	Pacific Developing Member Country
PAL	Polynesian Airlines
PE Act	Public Enterprise Act
PNCC	Palau National Communications Corporation
PNG	Papua New Guinea
PPL	Papua New Guinea Power Ltd
PPP	public–private partnership
PPP Act	Public–Private Partnership Act 2014 (Papua New Guinea)
PPUC	Palau Public Utilities Corporation
PSDI	Pacific Private Sector Development Initiative
PUB	Public Utilities Board
RMIPA	Republic of Marshall Islands Port Authority
ROA	return on assets
ROE	return on equity
SAL	Solomon Airlines
SCI	statement of corporate intent
SIEA	Solomon Islands Electricity Authority
SIPA	Solomon Islands Ports Authority
SIWA	Solomon Islands Water Authority
SOE	state-owned enterprise
SOEMAU	State-Owned Enterprise Monitoring and Advisory Unit
SOEMU	State-Owned Enterprise Monitoring Unit
SOI	Statements of Intent
SPX	South Pacific Stock Exchange
TA	technical assistance
TCC	Tonga Communications Corporation
TCFD	Task Force on Climate-Related Financial Disclosures
TCPA	Tobolar Copra Processing Authority
TERM	Tonga Energy Roadmap
TPL	Tonga Power Ltd
UNELCO	Union Electrique du Vanuatu
VADB	Vanuatu Agriculture Development Bank
VBTC	Vanuatu Broadcasting and Television Corporation
VFM	value for money
VPL	Vanuatu Post Limited

Currencies

In this report, "$" refers to United States dollars, unless otherwise stated.

A$	Australian dollar
F$	Fiji dollar
K	kina
NZ$	New Zealand dollar
SI$	Solomon Islands dollar
ST	tala
T$	pa'anga
Vt	vatu

Executive Summary

This study reviews the historical financial performance of state-owned enterprises (SOEs) in nine Pacific island economies,[1] identifies the risks that state-owned power utilities are facing as a result of climate change, and assesses the policy mechanisms that are available to governments and SOEs to improve their resiliency and sustainability. The study focuses on the period 2015–2022, thereby including the impact of the first 2 years of the coronavirus disease (COVID-19) and its associated border closures and lockdowns.

This is the seventh benchmarking study in the Finding Balance series, which aims to track the progress and impact of SOE commercialization in participating countries. This principle of commercialization, which allows SOEs to operate under similar legal and governance parameters as private companies, underpins SOE reform efforts throughout the Pacific, and is compatible with broader government goals of inclusive economic growth. Due to its focus on commercialization, the study does not detail the specific technical performance indicators of the SOEs.

In addition to the survey of SOE financial performance, this edition of the Finding Balance series has a special focus on the profound risks posed by climate change, and how governments and their state-owned utilities are acting to mitigate its effects and build resiliency. The study shows how efforts to commercialize SOEs and partner with the private sector are assisting SOEs to respond to climate change.

As the benefits of SOE commercialization have been detailed in previous editions of the Finding Balance series, they are not presented again here. The core finding of this research, however, bears repeating: the SOE model is not an effective long-term ownership structure. While the SOE model attempts to replicate private ownership incentives and dynamics, it never truly replaces the market disciplines that private firms face, since elected officials will tend to avoid commercial decisions with potential short-term political costs. Wherever possible, governments should seek to incorporate private sector participation into the provision of SOE services, so as to lock in the commercial incentives for improved performance.

Partnerships with the private sector, through full or partial privatization, supported by robust regulatory arrangements, are the most effective mechanisms for long-term improvement in state assets' productivity. Where full privatization is not politically feasible nor the most suitable reform mechanism, partial privatization (public listings, joint ventures, and public–private partnerships [PPPs]) can help to improve SOE performance.

---

[1]   Fiji, Kiribati, the Marshall Islands, Palau, Papua New Guinea, Samoa, Solomon Islands, Tonga, and Vanuatu.

The SOE portfolios of the nine Pacific Developing Member Countries (PacDMCs) surveyed in this study are dominated by infrastructure service providers (e.g., airports, seaports, power, water, sanitation, broadcasting, postal services, and telecommunications), and also include a range of other commercially oriented undertakings such as transport and banking.

The study shows that, while average SOE returns have improved over the past decade, they still fall far short of covering their costs of capital. Only two of the nine SOE portfolios produced a return sufficient to cover capital costs between 2015 and 2020. Three produced average returns on assets (ROAs) and/or average returns on equity (ROEs) below zero over this period.

Table 1: State-Owned Enterprise Portfolio Performance Indicators (%)

Country	Average Return on Assets FY2015–FY2020	Average Return on Equity FY2015–FY2020	Contribution to Gross Domestic Product 2020	Total Portfolio Assets 2020 ($m)
Fiji	3.4	7.4	2.6	2,002
Kiribati[a]	2.4	2.9	14.6	172
Marshall Islands	4.6	7.0	8.8	198
Palau	(1.2)	(2.7)	5.4	198
Papua New Guinea	(0.2)	(0.1)	1.2	2,866
Samoa	0.4	0.8	4.1	645
Solomon Islands	7.1	9.8	4.1	386
Tonga	2.7	4.8	5.8	304
Vanuatu	(2.2)	(21.4)	(0.1)	312

( ) = negative, FY = fiscal year

[a] The financial results of the Kiribati state-owned enterprise portfolio must be treated with some caution, as some of the state-owned enterprises received unqualified audit reports.

Sources: Asian Development Bank, ADB Key Indicators. https://kidb.adb.org; World Bank. World Bank Development Indicators. http://databank.worldbank.org/data/home.aspx; and state-owned enterprise accounts provided by countries and publicly available sources.

In most countries, these low returns are achieved despite subsidized capital, monopoly market power, and ongoing government cash transfers. The low returns on SOE investment dampen economic growth. Despite governments' sizeable investments in the SOEs, representing up to an estimated 34% of total fixed capital in each country, they contributed 0%–15% to gross domestic product in 2020.

Across the Pacific region, gross domestic product contracted 6.4% in 2020 and 1.5% in 2021,[2] largely due to travel restrictions which cut off a substantial portion of tourism inflows, labor mobility, and trade. This economic contraction was mirrored in the SOE portfolios of each country, with five of the nine countries recording lower returns in 2020 compared to 2019. In contrast, the Marshall Islands, Kiribati, and Palau saw improved portfolio performance because of a range of factors, including sharply reduced expenses in the power utilities sector in Palau and the Marshall Islands (with revenue remaining steady), an increase in community service obligation payments to Air Kiribati, and a surge in non-core revenue in Kiribati's Plant and Vehicle Unit.

SOEs involved in air travel and tourism—most notably airlines and airports—were the hardest hit by border closures. Revenues were down an average of 34% for airlines and 27% for airports in 2020 compared to 2019, which weakened portfolio returns in five of the seven countries with SOE airlines and all 5 countries with airport SOEs.[3] Five of the seven airlines in the SOE portfolios were financially vulnerable before 2020, with some generating losses in each of the 5 years preceding 2020; COVID-19 only exacerbated their need for restructuring. Power and water utilities, in contrast, did not face a substantial reduction in revenues or profitability, despite their contributions to tariff subsidies in several countries.

As governments in the Pacific work to improve the financial sustainability of their SOEs, they must incorporate strategies to address emerging risks, including climate change, which is rapidly becoming existential. Climate scientists expect equatorial regions like the Pacific will confront more intense and more frequent natural hazards, such as cyclones, floods, and coastal inundation, as global temperatures increase in the coming decades. While Pacific countries contribute very little to global greenhouse gas (GHG) emissions, they are setting ambitious emissions reduction goals, investing in adaptation, and making plans for the worst impacts of climate change.

State-owned utilities are important players in these efforts. In eight of the nine countries in this survey, state-owned utilities are the dominant providers of electricity, 78% of which on average in 2020 was generated from thermal fuel, resulting in a substantial share of the region's GHG emissions.[4] At the same time, some of these utilities are leading the transition to renewable energy. While this transition is primarily driven by falling renewable energy costs and a desire to improve fuel security, state-owned utilities are also seeking to support the decarbonization goals of their state shareholders.

---

2   https://data.adb.org/dataset/gdp-growth-asia-and-pacific-asian-development-outlook.

3   Core revenue for Airports Kiribati Authority (AKA) grew by 8% from 2019 to 2020; it was the only airport in the surveyed countries where core revenue increased in 2020. Source: AKA draft 2020 financial statements.

4   Figure is the average percentage of fuel sources used in generation in the eight surveyed utilities in 2020; the range was from 36% to 100%.

This paper posits that Pacific state-owned utilities that are more commercialized may be more inclined to respond to commercial incentives to decarbonize, or to reduce the financial risks associated with climate change. These SOEs may also be more inclined to invest in protecting their assets from climate-induced physical damage, given its impact on financial sustainability. However, to optimize these commercial choices, SOEs must operate in a policy environment conducive to commercialization.

Beyond establishing robust governance frameworks within which their SOEs can operate commercially, governments have a number of policy tools with which to assist their SOEs to adapt to climate risks and build resilience. These include: (i) introduce climate-related financial risk disclosures and support SOEs to make such disclosures, as is required under the Papua New Guinea (PNG) Climate Change (Management) Act 2015; (ii) raise capital to finance public investment in climate resilience and decarbonization, in particular through development partners, green bonds, and specialized funds; and (iii) support private investment in renewable energy generation, including in existing state-owned electric utilities.

The final section of the study profiles each of the nine participating countries, detailing (i) the composition and financial performance of their SOE portfolios; (ii) their legal, governance, and monitoring frameworks; (iii) SOE reform milestones achieved in 2015–2022; and (iv) how their state-owned electric utilities are impacted by and responding to climate change.

Key SOE reform milestones between 2015 and 2022 include:

Fiji: privatizing 59% of Fiji Ports Corporation Limited in 2016 and 24% of Energy Fiji Limited (EFL) in 2019, with a further divestment of 20% in 2021, and it is intended that 5% of EFL's shares will be listed on the South Pacific Stock Exchange and given to existing customers; EFL signing a power purchase agreement for the development of a 5-megawatt (MW) grid-connected solar plant in 2021; and enacting a new Public Enterprise Act in 2019, which improved the governance and commercial framework for SOEs.

Kiribati: completing a competitive tender for a public-private partnership (PPP) to rehabilitate the Betio Shipyard in 2017, placing the Captain Cook Hotel under a management contract, and building capacity in the SOE Monitoring Unit (SOEMU) to implement the SOE Act.

Marshall Islands: adopting a comprehensive SOE Act in 2015, and operationalizing a SOEMU in the Ministry of Financein 2018.

Palau: completing a competitive tender for a 16MW solar generation PPP in 2020; issuing Executive Order 465 in 2021 reaffirming Palau's National Policy for SOE Governance and supporting SOE commercialization; preparing an SOE bill to implement the policy; undertaking a series of corporate planning and governance reforms in the Palau Public Utilities Corporation in 2021, consistent with the SOE Policy; establishing an independent regulatory process for setting cost recovery electricity tariffs for the Palau Public Utilities Corporation; and adopting a new Public–Private Partnership Policy in 2021.

Papua New Guinea: adopting a new SOE Ownership and Reform Policy in 2020 and subsequent amendments to the Kumul Consolidated Holdings (KCH) Act in 2021 to strengthen governance and transparency; operationalizing the Public-Private Partnership Act through the adoption of amendments and regulations in 2022; completing a competitive tender for a container terminal concession for the ports of Lae and Port Moresby in 2017; signing two independent power producer (IPP) contracts for gas-fired power generation in 2019; and launching a competitive tender for a solar IPP in 2021.

Samoa: amending the SOE Act in 2015 to establish a new SOE ministry under a minister of SOEs; privatizing Agricultural Stores Corporation in 2016; and contracting four IPPs to add 10,000 kilowatts of solar generation capacity.

Solomon Islands: signing the power purchase agreement and reaching financial close for the Tina River Hydro Public–Private Partnership Project; corporatizing airport operations into the Solomon Islands Airport Corporation; issuing a new SOE Ownership Policy in 2018 to guide future investments, link the target capital structure and dividend of each SOE to its business plan, and strengthen the coherence of SOE oversight; and Cabinet endorsement of amendments to the SOE Act to strengthen the director selection process and promote women's participation on SOE boards.

Tonga: completing a competitive international tender for a 6MW solar IPP, which reached financial close in 2020 and was commissioned in 2022; completing a concession agreement for Tonga Forest Products Ltd assets and plantation on 'Eua Island; amending the SOE Act in 2020 and 2021 to strengthen governance and monitoring provisions; completion of a competitive tender for an integrated cargo handling concession at the Queen Salote International Wharf; and Cabinet endorsement of skills-based SOE director selection guidelines in 2020.

Vanuatu: preparing a new SOE bill in 2018 based on the 2013 SOE ownership policy, which calls for the commercialization of SOEs and establishment of a centralized ownership and monitoring function within the Ministry of Finance and Economic Management; and the further privatization of National Bank of Vanuatu in 2020, with the Ministry of Finance reducing its ownership share to 44%.

I. Introduction

This study reviews the historical financial performance of state-owned enterprises (SOEs) in nine Pacific island economies,[5] identifies the risks state-owned power utilities are facing as a result of climate change, and assesses the policy mechanisms available to government and SOEs to improve resilience and sustainability. The study focuses on the period 2015–2022, thereby including the impact of the first 2 years of the coronavirus disease (COVID-19) and its associated border closures and lockdowns. As of late 2022, most of the surveyed countries did not have 2021 financial accounts available for all of their SOEs, so 2020 was used as the most recent year. Delays in the finalization of SOE accounts are a persistent challenge in the surveyed countries, with SOEs often in violation of their statutory obligations to file audited returns within a set time period. These delays are linked to a range of factors, from weak financial management systems to backlogs with auditors general.

This is the seventh benchmarking study in the Finding Balance series, and all nine surveyed countries were also included in either the 2016 or 2019 editions. Since 2019, the benchmarking studies have included a thematic focus in addition to the general survey of portfolio financial performance and progress towards increasing SOE commercialization. In 2019, this special focus was on state-owned banks. This year, we look at how governments and their state-owned electric utilities are taking action to reduce greenhouse gas (GHG) emissions and build resiliency in the face of climate change.

While the nine countries included in the 2023 edition vary significantly in size, population, and growth rates (Table 2), they are reasonably comparable because of their SOE reform history and broadly similar SOE portfolio composition.

Table 2: Survey Country Economic Indicators

Country	Population (2020)	GDP ($m, 2020)	GDP per Capita ($)	GDP Growth Average (2015–2019)	GDP Growth[a] (2020)
Fiji	896,440	4,476	4,993	5%	(18%)
Kiribati	119,450	181	1,515	3%	3%
Marshall Islands	59,190	245	4,139	7%	2%
Palau	18,090	258	14,262	0%	(6%)
Papua New Guinea	8,947,030	24,668	2,757	9%	2%
Samoa	198,410	829	4,178	3%	(9%)
Solomon Islands	686,880	1,536	2,236	6%	(5%)
Tonga	105,700	491	4,645	8%	(3%)
Vanuatu	307,150	916	2,982	8%	(8%)

( ) = negative, GDP = gross domestic product, $m = $ million.
a   The growth rate is calculated based on nominal price and in local currency.
Source: ADB Key Indicators. https://kidb.adb.org.

In this study, SOE refers to public enterprises, commercial government business enterprises, and public trading bodies that are majority-owned by the state. All are corporatized and, with few exceptions, have a statutory obligation to operate profitably. Mutual financial institutions, such as insurance companies and provident funds, are excluded from this study as their equity is owned by their contributors, not the government.[6] Similarly, companies that are majority-owned by provident funds are excluded. Resource and petroleum SOEs held and managed outside of the SOE Monitoring Units (SOEMUs) have also been excluded. In the case of Fiji, however, all majority-owned SOEs are included in the portfolio financial results, even though only 13 are regulated by the Public Enterprise Act and monitored by the Department of Public Enterprises. Where the state owns less than 100% of an SOE, its results are prorated into the portfolio totals in the same percentage as the state's ownership. A detailed list of the SOEs included in the study is provided in the Appendix.

The financial performance analysis focuses on the period 2015–2020, although data from 2010–2014 and 2021 is also included where available. The analysis of SOE reforms undertaken in each country covers the period 2015–2022. The study was prepared with the active support of the electric utilities and ministries of finance or public enterprises or other monitoring agencies in each of the survey countries. Each monitoring agency provided audited or unaudited financial information on its SOEs and completed a questionnaire broadly describing its SOE monitoring practices, governance arrangements, and reform milestones. Each electric utility also completed a questionnaire outlining its efforts to increase renewable energy use and build resiliency to climate change. This information was then discussed with each agency for further clarification, before being assessed comparatively across the nine countries.

Given the extensive analysis of best practices and global trends in SOE governance and legal frameworks in previous Finding Balance editions, we do not cover these topics here. Readers are invited to review the 2016 paper[7] for evidence of the benefits of SOE commercialization, and the key features of effective legal, regulatory, and governance arrangements. In this paper, we highlight the key reform measures undertaken between 2015 and 2022 to strengthen the commercial frameworks within which SOEs operate in each surveyed country.

---

[6] PNG's Motor Vehicle Insurance Limited is included in the PNG portfolio since it is a limited liability company whose shares are 100% owned by the state, not its contributors. Similarly, Kiribati Insurance Corporation is included in the Kiribati portfolio since it is a statutory corporation whose equity is funded and owned by the state.

[7] https://www.adb.org/sites/default/files/publication/192946/finding-balance-2016-soe.pdf.

II. Economic Impact of the State-Owned Enterprise Portfolios

A. Portfolio Composition

The SOEs in this study are primarily engaged in two broad activities: (i) the delivery of core public infrastructure services such as airports, seaports, power, water, sanitation, broadcasting, postal services, and telecommunications; and (ii) a range of other commercially-oriented undertakings such as transport, banking, food processing, property development, tourism, housing, agriculture, oil, and gas.

In eight of the nine countries surveyed in this study, infrastructure SOEs dominate the portfolio, representing 43%–100% of total assets in 2020 (Figure 1). Vanuatu's portfolio is unusual in that it does not have a power or water utility; its only infrastructure SOEs are the airport and port.[8]

Figure 1: Composition of State-Owned Enterprise Portfolios, FY2020

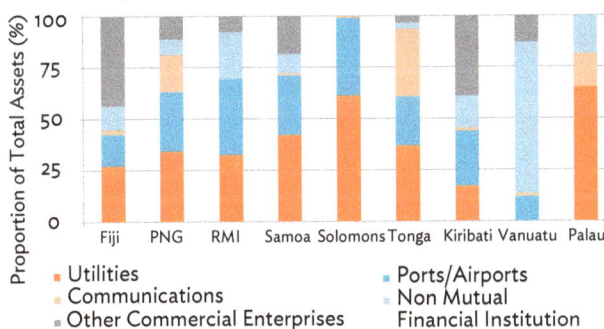

FY = fiscal year, PNG = Papua New Guinea, RMI = Republic of the Marshall Islands, SOE = state-owned enterprise.
Sources: Government of Fiji, Ministry of Economy; Government of Palau, Ministry of Finance and annual SOE audit reports; Government of PNG, Kumul Consolidated Holdings; Government of the Marshall Islands, annual economic statistics tables and annual SOE audit reports; Government of Samoa, SOE Monitoring Unit; Government of Solomon Islands, Ministry of Finance; Government of Tonga, Ministry of Public Enterprises; Government of Kiribati, SOE Monitoring and Advisory Unit, Ministry of Finance and Economic Development; and Government of Vanuatu, Department of Finance and Treasury.

B. Contribution to Gross Domestic Product

Investments in SOEs are substantial, yet their contribution to gross domestic product remains low. SOEs control 5%–37% of total fixed capital in each country, yet contributed only 0%–15% to gross domestic product (GDP) in 2020 (Figure 2).[9] In more mature economies such as New Zealand, SOEs controlled 2%–3% of total fixed capital and contributed 0.8% to GDP in 2020. The SOEs in this survey are among the largest commercial entities in their respective countries. In general, large businesses are almost always more productive than small businesses. When a SOE's contribution to GDP is much smaller than the portion of fixed assets they use in the economy, this inefficiency acts as a drag on economic growth.

On average, for every 1% share of total fixed capital, SOEs in the survey countries contribute 0.21% to GDP. By this measure, Kiribati is the best performer in the survey sample, with SOEs contributing 0.45% to GDP for each 1% share of total fixed capital, although this number is unaudited. New Zealand's SOEs contributed 0.36% to GDP for every 1% share of total fixed capital in 2020—almost double the average of the surveyed countries, excluding Kiribati.

---

[8] The 2020 accounts for Vanuatu Post were not available as of November 2022 and, therefore, are not included in this analysis; power and water in Vanuatu's major cities are provided via public-private partnership concession or via government department.

[9] Gross fixed capital formation data is not available for Samoa. Because of data deficiencies, there is a large margin for error in these calculations. However, even using the most optimistic estimates in a sensitivity analysis of the capital output ratio for sampled countries over a 10-year period, it appears the low productivity of state-owned enterprises could have resulted in a 10%–20% reduction in gross domestic product. This is a substantial economic cost imposed on this study's sample countries.

Figure 2: State-Owned Enterprise Contribution to Gross Domestic Product versus Total Fixed Capital Controlled by State-Owned Enterprises, FY2020

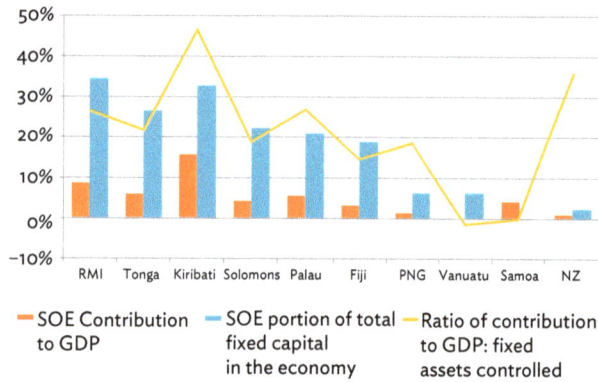

FY = fiscal year, GDP = gross domestic product, NZ = New Zealand, SOE = state-owned enterprise, PNG = Papua New Guinea, RMI = Republic of the Marshall Islands.
Sources: Asian Development Bank. 2021. Key Indicators for Asia and the Pacific 2021. Manila; Asian Development Bank estimates; World Bank. World Development Indicators. http://databank.worldbank. org/data/home.aspx; Government of Kiribati, SOE Monitoring and Advisory Unit, Ministry of Finance and Economic Development; Government of Tonga, Ministry of Public Enterprises; Government of the Marshall Islands annual economic statistics tables, and annual SOE audit reports; Government of Fiji, Ministry of Economy; Government of Solomon Islands, Ministry of Finance; Government of Samoa, SOE Monitoring Unit; Government of Vanuatu, Department of Finance and Treasury; Government of PNG, Kumul Consolidated Holdings; and Government of Palau, Ministry of Finance.

C. Impact on the Private Sector

In most of the countries in this study, vital infrastructure services, such as utilities and transport, are provided exclusively by SOEs. When these are of low quality and/or high cost, they affect the sustainability of the private sector.

SOEs reduce growth by crowding out the private sector and dampening the competitiveness of domestic industries. When SOEs compete with private sector companies, they often do so on a favored basis, complicating private sector competitors' efforts to invest and grow. Although private sector firms are generally more efficient, and are not burdened with community service obligations (CSOs), SOEs enjoy a competitive advantage in three key areas:

(i) Preferred access to government contracts.
(ii) Subsidized capital. SOE debt and equity costs are generally lower than those of private sector firms, allowing them to remain marginally profitable despite being less efficient than their private competitors.
(iii) Monopoly provision of services in some cases.

Subsidized debt, like subsidized equity, creates economic distortions. The interest rates that SOEs pay on their debt are substantially below commercial rates (Figure 3). The low financing costs are a result of

(i) explicit and implicit government guarantees; and
(ii) soft loans provided by government entities, or onlent from donors.

Figure 3: Average Cost of State-Owned Enterprise Debt versus Commercial Debt Rate, FY2015–FY2020

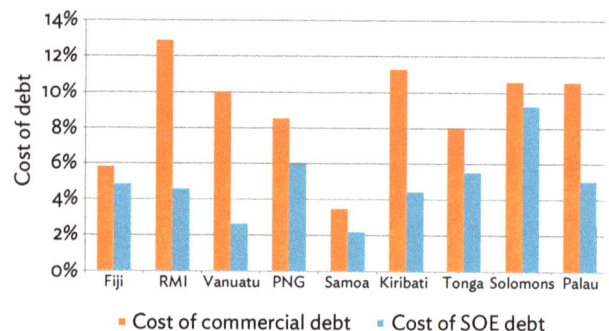

FY = fiscal year, PNG = Papua New Guinea, RMI = Republic of the Marshall Islands, SOE = state-owned enterprise
Sources: International Monetary Fund. International Financial Statistics; Government of Fiji, Ministry of Economy; Government of Vanuatu, Department of Finance and Treasury; Government of Samoa, SOE Monitoring Unit; Government of Tonga, Ministry of Public Enterprises; Government of PNG, Kumul Consolidated Holdings; Government of Kiribati, SOE Monitoring and Advisory Unit, Ministry of Finance and Economic Development; Government of the Marshall Islands, annual economic statistics tables and annual SOE audit reports; Government of Solomon Islands, Ministry of Finance; and Government of Palau, Ministry of Finance.

Box 1: Implementing Community Service Obligation Frameworks in the Pacific

Most state-owned enterprise (SOE) acts of Pacific island countries incorporate community service obligation (CSO) frameworks, which require agreed CSOs to be documented in a performance-based agreement between the SOE and the government. While the frameworks do not require the government to make a payment to SOEs to offset the costs of CSOs, all frameworks require CSOs to be identified and costed so that their financial impact on SOEs is disclosed. Ultimately, all agreed CSOs should be fully funded so that the SOEs can comply with their statutory requirements to operate commercially.

However, the implementation of frameworks has varied, which often depends on the technical capacity of the SOEs and monitoring agencies, the availability of CSO funding, and political considerations. For example:

(i)   In **Fiji**, Energy Fiji Ltd has a CSO contract with the government for the supply of electricity to rural communities, and receives a payment each year based on actual net cost of delivery.
(ii) In **Tonga**, three SOEs provided CSOs in 2020, all of which were costed and funded.
(iii) In **Solomon Islands**, SOEs consistently submit CSO costings, but not all are fully funded. Consideration for the CSO provided by Solomon Islands Water Authority in 2020, for example, was partially funded and partially acknowledged through a reduced return on equity target.
(iv) In **Kiribati**, Air Kiribati received CSO payments of A$19.1 million in the period 2018–2020. Its CSO framework was only partially implemented.
(v)   In the **Marshall Islands**, three SOEs—Tobolar Copra Processing Authority, Marshall Islands Shipping Corporation, and Air Marshall Islands—received CSO payments totaling $11.6 million in 2020, despite the CSO framework being only partially implemented.
(vi) In **Papua New Guinea**, a new CSO framework was incorporated into the amendments to the Kumul Consolidated Holdings Act in 2021, and implementation began with the identification of CSOs in 2022. Historically, SOEs such as Air Niugini have managed CSO services directly with "purchasers", such as provincial governments, which have financed new routes. Other SOEs, such as PNG Power and PNG Ports, have funded CSOs through cross-subsidization.

Source: Pacific Private Sector Development Initiative.

D. Fiscal Impact

Infrastructure SOEs are often forced to provide services on noncommercial terms. This typically involves delivering services to remote populations or providing services at reduced prices to selected customer groups. If properly costed, contracted, and funded, delivering these CSOs should not reduce the SOEs' profitability. While the implementation of CSO frameworks is improving in the Pacific, they remain generally underfunded (Box 1). These CSOs depress SOE profitability, contribute to inefficient resource allocation, and impair the government's ability to assess CSO value for money (VFM) or outcome achievements.

SOEs benefit from ongoing government equity contributions. These are typically provided to finance assets and working capital, retire debt, and absorb accumulated losses to allow SOEs to function. Seven of the nine countries in the survey received ongoing government support (Figure 4), contributing to fiscal deficits.

SOE net profits exceeded the value of government transfers in only four countries in 2015–2020.[10] The majority of countries with cumulative net losses between 2015 and 2020 require ongoing government contributions. For countries in which ongoing government contributions are required, these are equivalent to 0.02%–3.74% of GDP (Figure 5).

---

[10]   Fiji, the Marshall Islands, Solomon Islands, and Tonga.

Figure 4: Total Government Transfers to State-Owned Enterprises versus Total State-Owned Enterprise Net Profits, FY2015–FY2020

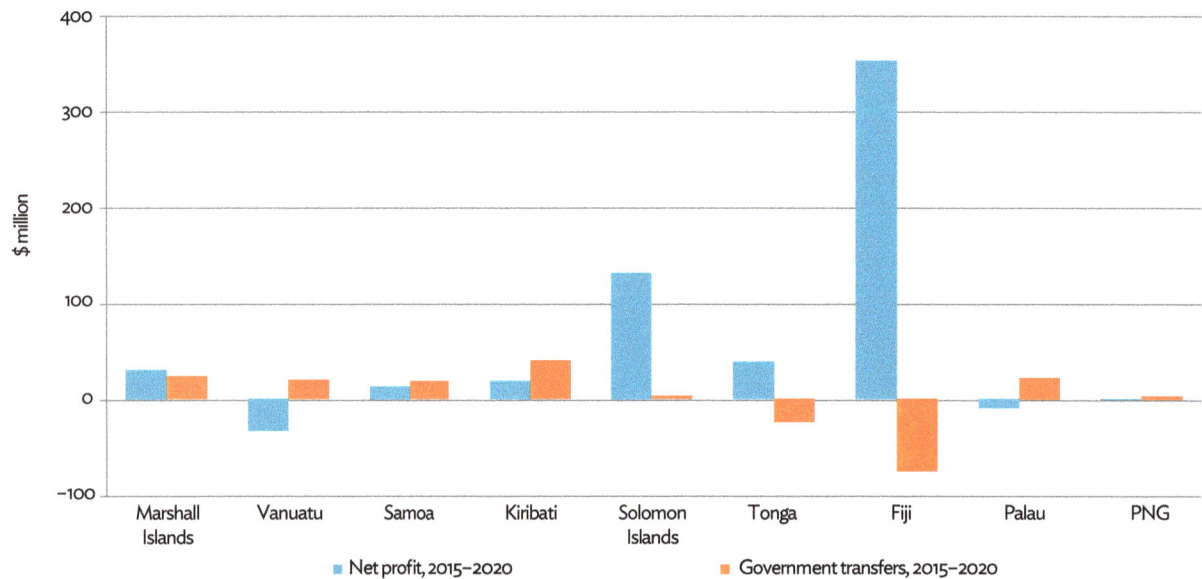

FY = fiscal year, PNG = Papua New Guinea, SOE = state-owned enterprise.
Sources: Government of Fiji, Ministry of Economy; Government of Tonga, Ministry of Public Enterprises; Government of Solomon Islands, Ministry of Finance; Government of Kiribati, SOE Monitoring and Advisory Unit, Ministry of Finance and Economic Development; Government of Vanuatu, Department of Finance and Treasury; Government of Samoa, SOE Monitoring Unit; Government of the Marshall Islands, annual SOE audit reports; Government of PNG, Kumul Consolidated Holdings; and Government ofPalau, Ministry of Finance.

Figure 5: Average Government Transfers to State-Owned Enterprises as a Percentage of Average Gross Domestic Product, FY2015–FY2020

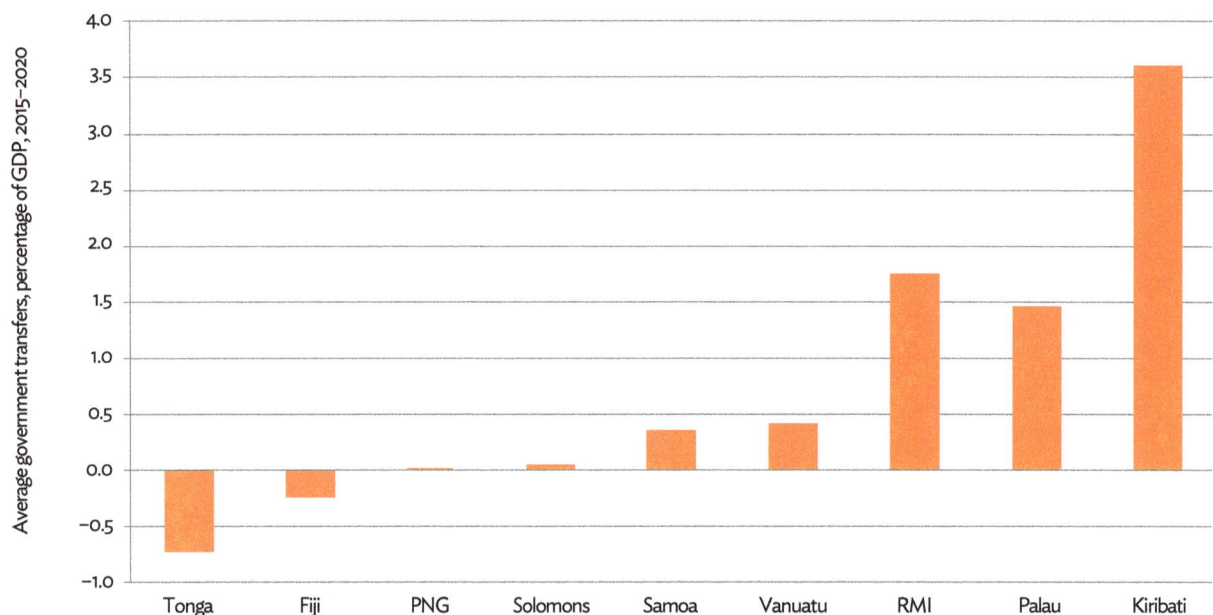

FY = fiscal year, GDP = gross domestic product, PNG = Papua New Guinea, RMI = Republic of the Marshall Islands, SOE = state-owned enterprise.
Sources: World Bank. World Development Indicators. http://databank.worldbank.org/data/home.aspx; Government of Kiribati, SOE Monitoring and Advisory Unit, Ministry of Finance and Economic Development; Government of Tonga, Ministry of Public Enterprises; Government of Solomon Islands, Ministry of Finance; Government of Fiji, Ministry of Economy; Government of PNG, Kumul Consolidated Holdings; Government of Palau, Ministry of Finance; Government of Vanuatu, Department of Finance and Treasury; Government of Samoa, SOE Monitoring Unit; and Government of the Marshall Islands, annual economic statistics tables and annual SOE audit reports.

E. State-Owned Enterprise Financial Performance, 2015–2020

The financial performance of most SOE portfolios is weak.
While the average financial performance of four of the SOE portfolios improved in 2015–2020 compared with the 2010–2014 period, the average return on assets (ROA) and return on equity (ROE)[11] remain below the returns that private sector investors would expect for comparable business risk (Figures 6 and 7, and Table 3). The exception is Solomon Islands, which averaged a nominal 9.8% ROE and 7.1% ROA in 2015–2020. Fiji and the Marshall Islands markedly improved their average portfolio returns in 2015–2020 as compared to 2010–2014, with Fiji doubling its portfolio ROE and ROA, and the Marshall Islands moving from negative returns in 2010–2014 to a nominal ROE of 7% and ROA of 4.6% in 2015–2020. If inflation is factored in to calculate real ROE and ROA, average profitability is even lower. Inflation, as measured by the GDP deflator,[12] averaged between 1.3% and 4.2% per year over the FY2015–FY2020 period in the surveyed countries (Figure 6 and Table 3).

Figure 6: Average Inflation Rate and Nominal Returns on Equity and Assets of State-Owned Enterprise Portfolios, FY2015–FY2020

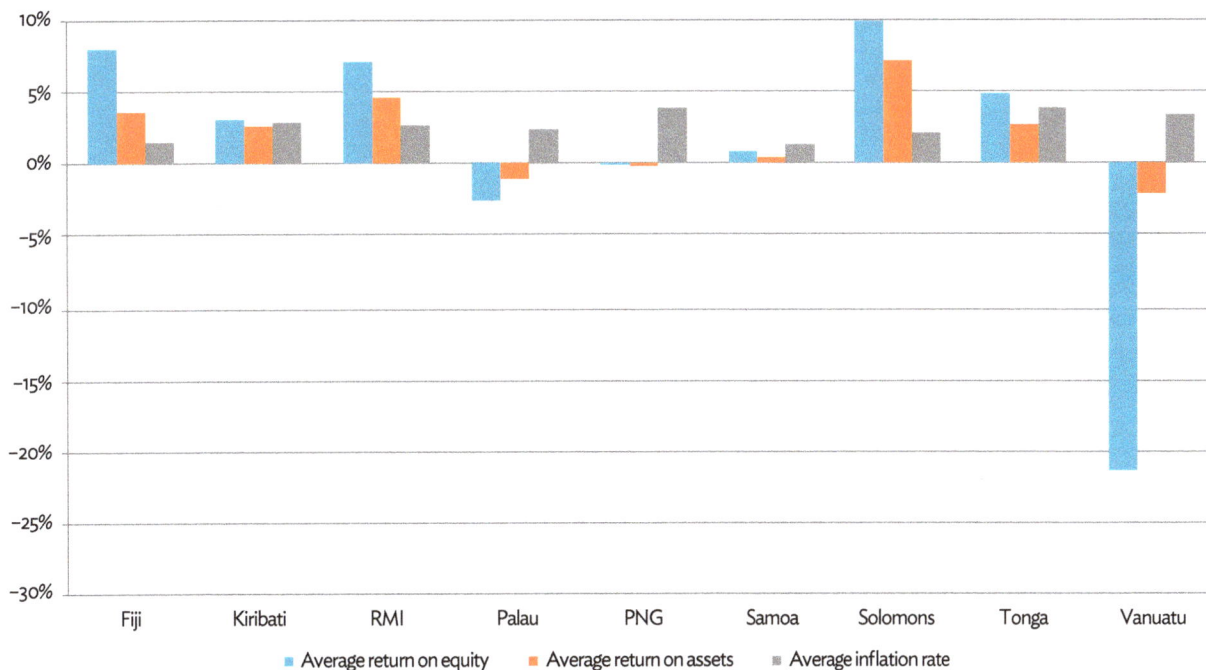

FY = fiscal year, PNG = Papua New Guinea, RMI = Republic of the Marshall Islands, SOE = state-owned enterprise.
Sources: Government of Fiji, Ministry of Economy; Government of Tonga, Ministry of Public Enterprises; Government of Solomon Islands, Ministry of Finance; Government of Kiribati, SOE Monitoring and Advisory Unit, Ministry of Finance and Economic Development; Government of Vanuatu, Department of Finance and Treasury; Government of Samoa, SOE Monitoring Unit; Government of the Marshall Islands SOE audit reports; Government of PNG, Kumul Consolidated Holdings; Government of Palau, Ministry of Finance; and World Bank Open Data https://data.worldbank.org/.

---

[11] Return on assets and return on equity are important indicators of how efficiently state-owned enterprises use their capital resources, but differ depending on how much debt is used to finance operations.

[12] The annual growth rate of the gross domestic product implicit deflator shows the rate of price change in the economy as a whole.

Table 3: State-Owned Enterprise Portfolio Profitability Indicators

Country	Average Nominal Return on Assets (%)			Average Nominal Return on Equity (%)			Average Inflation Rate (%)
	FY10–FY14	FY15–FY20	FY10–FY20	FY10–FY14	FY15–FY20	FY10–FY20	FY15–FY20
Fiji	1.7	3.4	2.6	3.7	7.4	5.7	1.5
Kiribati	2.8	2.4	2.6	3.7	2.9	3.3	2.6
Marshall Islands	(2.6)	4.6	1.3	(5.8)	7.0	1.2	2.7
Palau	2.2	(1.2)	0.4	(4.8)	(2.7)	(3.0)	2.4
Papua New Guinea	0.9	(0.2)	0.3	1.5	(0.1)	0.6	4.2
Samoa	(0.2)	0.4	0.1	(0.3)	0.8	0.3	1.3
Solomon Islands	7.1	7.1	7.1	10.5	9.8	10.1	2.1
Tonga	2.6	2.7	2.6	3.9	4.8	4.4	3.9
Vanuatu	0.0	(2.2)	(1.2)	0.4	(21.4)	(11.5)	3.4

( ) = negative, FY = fiscal year, SOE = state-owned enterprise.
Note: Return on assets is calculated as net profit after tax over total assets. Return on equity is calculated as net profit after tax over total equity. Asset utilization is calculated as total sales over total asset. All numbers are nominal. Inflation is the gross domestic product deflator.
Sources: Government of Tonga, Ministry of Public Enterprises; Government of Papua New Guinea, Kumul Consolidated Holdings; Government of Kiribati, SOE Monitoring and Advisory Unit, Ministry of Finance and Economic Development; Government of the Marshall Islands, annual economic statistics tables and annual SOE audit reports; Government of Fiji, Ministry of Economy; Government of Solomon Islands, Ministry of Finance; Government of Samoa, SOE Monitoring Unit; Government of Vanuatu, Department of Finance and Treasury; Government of Palau, Ministry of Finance; and World Bank Open Data https://data.worldbank.org/.

Most surveyed countries saw their average SOE portfolio ROA improve in 2016 and 2017 before falling back to 2015 levels by 2020. Apart from the Marshall Islands and Kiribati, most countries were unable to materially improve their ROA after 2018 (Figure 7). It is notable that, despite a downward trend, Solomon Islands was able to maintain a healthy average nominal ROA of 7.1% for the 2015–2020 period, due in part to cost-based tariffs in its electricity SOE, a doubling of port tariffs in 2016, and annual CSO payments which offset a portion of CSO costs.

Figure 7: Average Nominal Return on Assets of State-Owned Enterprise Portfolios, FY2015–FY2020

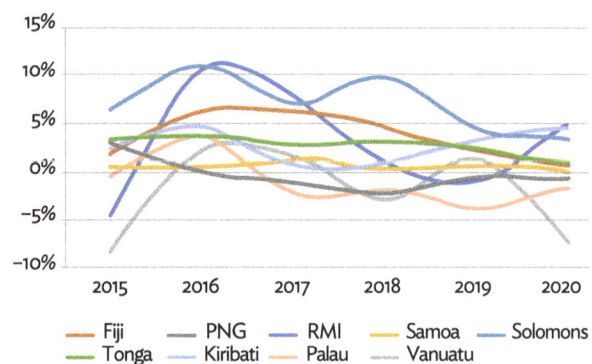

FY = fiscal year, PNG = Papua New Guinea, RMI = Republic of the Marshall Islands, SOE = state-owned enterprise.
Sources: Government of Tonga, Ministry of Public Enterprises; Government of PNG, Kumul Consolidated Holdings; Government of Kiribati, SOE Monitoring and Advisory Unit, Ministry of Finance and Economic Development; Government of the Marshall Islands, annual SOE audit reports; Government of Fiji, Ministry of Economy; Government of Solomon Islands, Ministry of Finance; Government of Samoa, SOE Monitoring Unit; Government of Vanuatu, Department of Finance and Treasury; and Government of Palau, Ministry of Finance.

F. Impact of COVID-19

Economically, the Pacific region contracted 6.4% in 2020 and 1.5% in 2021 largely because of COVID-19 travel restrictions, which cut off a substantial portion of tourism inflows, labor mobility, and trade. This economic contraction was mirrored in the SOE portfolios of each country, with SOEs involved in air travel and tourism—most notably airlines and airports—hardest hit. Revenues were down an average 34% for airlines and 27% for airports in 2020 as compared to 2019, weakening portfolio returns in five of the seven countries with SOE airlines and all five countries with SOE airports (footnote 3). Fiji and Vanuatu were the hardest hit, with their airlines suffering a 68% drop in revenue in 2020, while their airport revenues dropped 65% and 52%, respectively, from 2019 to 2020. With the border closures persisting through 2021, and in some cases 2022, these SOEs are likely to suffer reduced revenues well into 2023. Fiji Airways reported a further 37% decline in revenue in 2021.

Figure 8: Average State-Owned Enterprise Return on Assets by Sector in Surveyed Countries, FY2015–FY2020

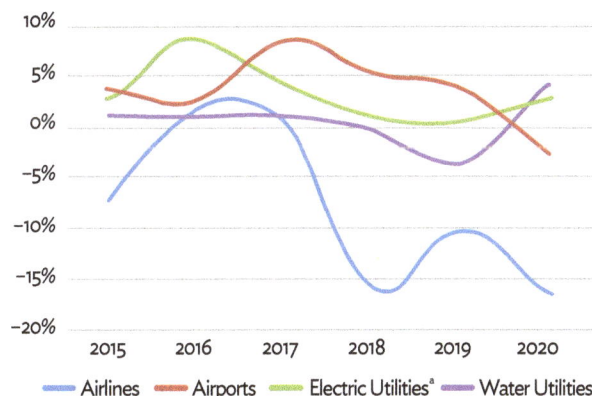

FY = fiscal year, SOE = state-owned enterprise.
a   The electric utilities include the Palau Public Utilities Corporation and the Public Utilities Board (Kiribati), which also provide water services. They have not been counted in the water utilities group.
Sources: Marshall Islands SOE audit reports; Government of Samoa, SOE Monitoring Unit; Government of Solomon Islands, Ministry of Finance; Government of Tonga, Ministry of Public Enterprises; Government of Kiribati, SOE Monitoring and Advisory Unit, Ministry of Finance and Economic Development; and Government of Vanuatu, Department of Finance and Treasury.

Five of the seven airlines included in this study experienced a sharp drop in profitability. Air Vanuatu and Polynesian Air suffered the largest downturns, with a –53% and –33% ROA, respectively, in 2020. In contrast, Air Kiribati and Air Marshall Islands (AMI) improved their financial performance in 2020, compared to 2019, largely because of demand for cargo and repatriation flights, combined with a sharp drop in fuel prices. These fuel prices rose again in 2021, contributing to reduced profitability for AMI. In 2021, AMI's total operating expenses grew by 21%, outpacing growth in operating revenue (16%), and driving down ROE from 11% to 0%.

As a group, power and water SOEs, which in many cases shouldered some of the costs of government responses to COVID-19, managed to improve their profitability in 2020. For the eight power utilities, most of which are large consumers of diesel fuel, the drop in fuel prices was a major driver of reduced operating costs. Only Tonga Power, Solomon Islands Electricity Authority (SIEA), and Electric Power Corporation (EPC) Samoa experienced a drop in profitability in 2020 compared to 2019. For the four power utilities which had 2021 financial results available,[13] ROA in 2021 was similar to 2020. For the six water utilities, demand remained strong in 2020 and all experienced improved profitability, except for Tonga Water Board. In our survey sample of eight electric utilities, only two (Palau Public Utilities Corporation [PPUC] and Kiribati's Public Utilities Board) also provide water services. While PPUC is prohibited by law from cross-subsidizing between these two service lines, Public Utilities Board does not have the same restriction.

13   Electricity Fiji Ltd, Electric Power Corporation (Samoa), Tonga Power Limited, and Solomon Islands Electricity Authority (SIEA).

Overall, four of the nine SOE portfolios surveyed underwent a deterioration in portfolio ROA in 2020 compared to 2019. Kiribati, the Marshall Islands, Papua New Guinea (PNG), and Palau managed to improve portfolio ROA; in Kiribati and the Marshall Islands, this may be because their economies did not contract in 2020, while Palau's portfolio does not include airlines or airports.

SOEs implementing investment projects that incorporated foreign expertise and imported components experienced substantial delays because of border closures and supply chain disruptions. While these projects are expected to be completed, the future investment pipelines of these SOEs—airlines and airports in particular—are likely to be impacted by their weakened financial positions.

III. State-Owned Enterprise Reform Highlights, 2015–2022

A. Policy and Legal Reforms

Seven of the nine countries in this benchmarking study have SOE acts, and the remaining 2 have SOE bills they intend to table in their respective legislatures in 2023. While a robust SOE legal framework does not guarantee improved SOE performance, as demonstrated by the countries in this study, it does facilitate commercial outcomes where the political will exists for its implementation. Where the political will does not exist and SOE laws are ignored or their core provisions repealed, poor performance becomes acceptable and positive change is resisted.

Eight of the nine countries in this benchmarking survey adopted a new policy, law, or amended law to support SOE commercialization in the 2015–2022 period. The policies and laws include key provisions establishing the government monitoring function and requiring SOEs to operate profitably, prepare corporate plans, appoint directors based on required skills, account for CSOs, and publish annual accounts in a timely manner. In the case of the Marshall Islands, the 2015 SOE Act was amended in 2016 to repeal a provision restricting the appointment of ministers and public officials to SOE boards, thereby weakening a key governance principle. In PNG, the 2015 Kumul Consolidated Holdings Act was amended in 2021 to restore key SOE oversight powers to Kumul Consolidated Holdings, the SOE holding company, and remove them from Cabinet, which was ill-suited for the role. In Samoa, amendments to the Public Bodies Performance and Accountability Act in 2015 established the Ministry of Public Enterprises as the SOE ownership monitor, shifting the responsibility from the Ministry of Finance. In Fiji, a new Public Enterprise Act (PE Act) was passed in 2019, replacing the outdated 1996 act and strengthening governance and oversight provisions. However, the act only applies to 13 of the 20 active SOEs in Fiji's portfolio. In Palau and Vanuatu, SOE bills were prepared for the first time to strengthen governance and transparency and introduce an ownership monitoring function, but both are still pending parliamentary consideration.

Table 4: State-Owned Enterprise Policy and Legal Reforms, 2015–2022

Country	New SOE Policy	New or Amended SOE Legislation
Fiji		Public Enterprise Act 2019, improving governance and commercial provisions
Marshall Islands		SOE Act 2015
Palau	Yes	SOE bill prepared
Papua New Guinea	Yes	Kumul Consolidated Holdings Authorisation (Amendment) Act 2021, strengthening governance and transparency provisions
Samoa		Amended Public Bodies Performance and Accountability Act 2015 to establish the Ministry of Public Enterprises
Solomon Islands	Yes	SOE Amendment bill prepared
Tonga		Amended Public Enterprise Act in 2020 and 2021 to strengthen governance and monitoring provisions
Vanuatu		SOE bill prepared, based on 2013 SOE Ownership Policy
Kiribati		Amended SOE Act in 2016 to allow one sector ministry employee to be appointed to the related SOE board.

SOE = state-owned enterprise
Source: Pacific Private Sector Development Initiative.

B. State-Owned Enterprise Director Selection and Board Composition

The 2016 edition of Finding Balance observed strong evidence of a correlation between board composition and SOE performance in the Pacific. The director selection process is critical to establishing effective boards. One of the greatest challenges in the SOE board nomination process is balancing the level of political involvement with the need to ensure that the board is commercially independent. As government owns the SOE, government has a legitimate role in appointing directors. However, the selection of preferred candidates is best left to non-political agencies, such as the SOE ownership monitor, an independent selection committee, or a professional search firm reporting to the ownership monitor.

Government's primary responsibility is to establish a codified process that creates the highest probability that the best-qualified candidate will be selected. This process is anchored in an objective analysis of the skills, experience, independence, and diversity needed on each board, which then informs the terms of reference for each new vacancy, and the evaluation of all candidates. Seven of the nine countries surveyed in this study have an overarching SOE act that includes provisions on the qualification, disqualification, and nomination of directors, and six of these seven prohibit the appointment of elected officials and limit the number of appointed public servants. PNG, Samoa, Solomon Islands, and Tonga have codified their director selection and appointment processes, further reducing the risk of suboptimal or politicized appointments.[14]

All countries participating in this study have indicated a desire to increase the percentage of women directors on SOE boards. While only PNG and Samoa have set formal targets (30% female directors in PNG, at least one woman director per SOE in Samoa), the remainder recognize the value of increased gender diversity for improved board functioning and decision-making. Despite this consensus, only three of the seven surveyed countries for which data is available increased the percentage of female directors on SOE boards in 2021 compared to 2015 (Table 5). In PNG, a database of eligible women SOE directors has been developed, with efforts being made to provide them access to training and mentoring.

Table 5: State-Owned Enterprise Board Composition, 2021[a]

	Fiji	Marshall Islands	Kiribati	Palau	PNG	Samoa	Solomon Islands	Tonga	Vanuatu
Number of SOEs	24	11	18	4	10	15	9	12	7
Number of directors	111	85	91	18	63	74	50	30	38
Female directors (%)	20	26	25	39	8	27	10	10	0
Female directors in 2015 (%)	5	24	25	n/a	16	21	13	16	5
Number of elected officials serving as directors	0	25	0	2	0	0	0	0	2
Number of public servants serving as directors	22	37	53	1	0	1	4	0	2
Elected officials/public servants on boards (%)	20	73	58	17	0	0.1	8	0	11
SOEs with female chairs (%)	17	0	11	0	0	13	0	8	0
Average SOE Portfolio ROA 2015–2020 (%)	3.8	4.6	2.6	(1.2)	(0.3)	0.4	6.9	2.7	(2.2)

( ) = negative, PNG = Papua New Guinea, ROA = return on assets, SOE = state-owned enterprise.

a   Tonga has shared boards for SOEs grouped within a sector. For example, the same directors serve on the boards of Tonga Power Limited, Tonga Water Board, and Waste Authority Ltd, hence the small number of directors relative to the number of SOEs. In the Marshall Islands, SOEs in the utilities sector share the same directors.

Sources: Government of Tonga, Ministry of Public Enterprises; Government of PNG, Kumul Consolidated Holdings; Government of Kiribati, SOE Monitoring and Advisory Unit, Ministry of Finance and Economic Development; Government of the Marshall Islands, Ministry of Finance; Government of Fiji, Ministry of Economy; Government of Solomon Islands, Ministry of Finance and Treasury; Government of Samoa, Ministry of Public Enterprises; Government of Vanuatu, Department of Finance and Treasury; and Government of Palau, Ministry of Finance.

---

14   https://pacificpsdi.org/assets/Uploads/PSDI-SOE-Brief-Web.pdf.

C. Partnerships with the Private Sector

To cement the gains of SOE commercialization, governments have continued to divest some of their shareholdings and contract with the private sector for selected services. Where full privatization is not politically or practically feasible, provident funds have played a role in taking up shareholdings as a bridge to, or in tandem with, private sector investment. This has been the case in Fiji and Vanuatu, for example. Public-private partnership (PPP) contracts have also been used to provide selected services, in particular power generation, with seven new contracts signed during the 2015–2022 period.

PPPs can and have generated real benefits, and simple contracts are not costly. Countries should continue exploring PPP opportunities in their SOE portfolios so that bankable PPP projects can be implemented, and new opportunities identified. Critical to the success of PPPs, or joint ventures with SOE participation, are robust governance arrangements, full transparency, and arms-length relationships with governments—consistent with SOEs' commercial mandates.

Table 6: Selected Privatizations and Public–Private Partnerships, 2015–2022

Country	Privatizations and Public-Private Partnerships
Fiji	Privatization of 59% of Fiji Ports Corporation Privatization of 44% of Energy Fiji Ltd PPP for the development, financing, operation, and maintenance of the Lautoka and Ba hospitals Power purchase agreement signed for a 5MW solar IPP
Kiribati	Betio Shipyard PPP Captain Cook Hotel management contract
Palau	Completed tender for 16MW solar IPP
Papua New Guinea	Container terminal PPP for Lae and Port Moresby Ports Two gas-fired power generation IPP contracts Launched competitive tender for solar generation IPP Prepared PPP transaction for Jackson's Airport
Samoa	Four IPP contracts for power generation
Solomon Islands	Financial close reached for Tina River hydro IPP contract
Tonga	Completed tender for 6MW solar IPP Concession agreement for Tonga Forest Products
Vanuatu	Further privatization of National Bank of Vanuatu, with the Ministry of Finance reducing its ownership share to 44%

IPP = independent power producer, PPP = public–private partnership, MW = megawatt.
Source: Pacific Private Sector Development Initiative.

IV. Climate Change and State-Owned Electric Utilities

Climate change is one of the most significant economic, social, and political challenges facing Pacific Developing Member Countries (PacDMCs). Caused by increasing GHG emissions which in turn contribute to increasing global average temperatures, climate change will bring instability to the long-term weather conditions which underpin human societies and economies. Equatorial regions like the Pacific face particularly significant vulnerabilities from climate change, as scientists expect more intense and more frequent natural hazards, such as cyclones, floods, and coastal inundation. These impacts are compounded by the economic development challenges facing the region. At the same time, PacDMCs play only a minor role in the emissions underlying global warming, giving the region limited means to control the cause of its climate risks.

PacDMCs have taken steps to adapt to climate change across many dimensions of society. Governments are setting ambitious emissions reductions goals, investing in adaptation, and making plans for the worst impacts of climate change in the region. In this study, we consider one important, but often overlooked, dimension of such preparations: the way the main electricity sector actors in PacDMCs—SOEs—are responding to climate change.

Electricity sector SOEs have an important role to play in both emissions reduction, and in resilience against climate risks. In this section, we first consider how electricity systems impact and are impacted by climate change, and then evaluate how state-owned electric utilities in the Pacific are responding to, and could better respond to, climate risks facing the sector.

A. Electricity Systems, Greenhouse Gas Emissions, and Renewable Energy Targets

Electricity systems and SOEs are major drivers of climate change. Climate change is caused by an increase in GHG emissions in the atmosphere, which, since the Industrial Revolution, has been driven primarily by human combustion of fossil fuels. One of the primary uses of such fossil fuels is in electricity generation. While the use of renewable energy is increasing, totaling almost 29% of the source of electricity generation in 2020,[15] the sector remains one of the largest contributors to global GHG emissions.[16] Estimates indicate that most installed capacity of global electricity generation is carried out by electric utilities which are wholly or partially state-owned.[17]

In addition to their contribution to the greenhouse effect, electricity systems face a multitude of risks as a consequence of climate change. These risks include the physical impacts of climate change on energy infrastructure, such as curtailments in generation because of prolonged periods of high heat, physical damage to transmission infrastructure from wind, and generator infrastructure damage from coastal inundation.[18] Electric utilities also face risks associated with economic, social, and political changes prompted by climate change. For instance, utilities may find it harder to raise finance for electricity generation by high-emitting assets, as public and private lenders seek to reduce their exposure to these types of assets.[19]

Overall, the Pacific region contributes very little to global GHG emissions. Nonetheless, GHG emissions from the energy sector are increasing in the Pacific.[20]

---

[15]   International Energy Agency. 2021. *Global Energy Review 2021*. Paris.

[16]   Intergovernmental Panel on Climate Change. 2022. *Climate Change 2022: Impacts, Adaptation, and Vulnerability. Contribution of Working Group II to the Sixth Assessment Report of the Intergovernmental Panel on Climate Change*. Cambridge University Press, Cambridge, UK and New York, NY, USA.

[17]   A. Prag, D. Röttgers, and I. Scherrer. 2018. State-Owned Enterprises and the Low-Carbon Transition. *OECD Environment Working Papers, No. 129*. Paris. https://doi.org/10.1787/06ff826b-en.

[18]   Asian Development Bank (ADB). 2012. *Climate Risk and Adaptation in the Electric Power Sector*. Manila.

[19]   A. Gerlak et al. 2018. Climate risk management and the electricity sector. *Climate Risk Management*. 19. pp. 12–22.

[20]   ADB. 2020. *Pacific Energy Update 2020*. Manila.

Emissions intensity in the Pacific—the quantity of emissions produced to generate a kilowatt-hour of power—is also comparatively high. This is because the electricity generation mix across the Pacific relies on imported fossil fuels for more than half of its power demand, and power technologies are inefficient.[21] Figure 9 illustrates emissions per capita in the region compared to other economies.

Reliance on imported fuel contributes to high electricity tariffs in the Pacific. It also exposes Pacific countries to potential insecurity of supply and energy price volatility, as the cost of such fuels fluctuates in global energy markets. As illustrated in Figure 10, Pacific electricity tariffs exceed the world average.

Figure 9: Carbon Dioxide Emissions per Capita in Pacific Island Countries Compared to World Average and Other Major Economies, 1960–2018[a]

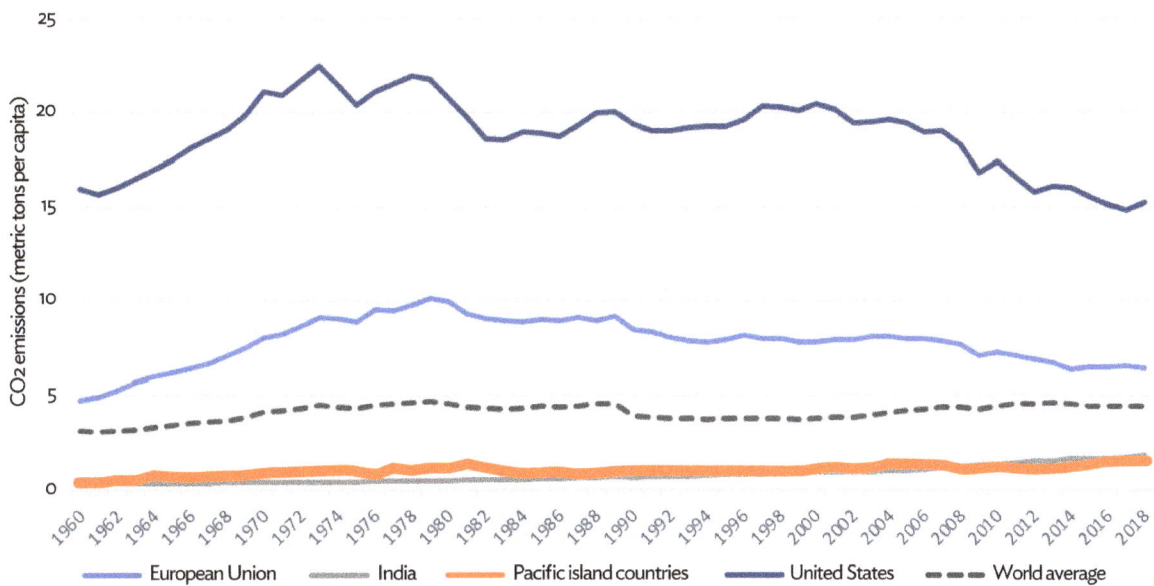

a   Carbon dioxide emissions are those stemming from the burning of fossil fuels and the manufacture of cement. They include carbon dioxide produced during consumption of solid, liquid, and gas fuels, and gas flaring.
Sources: Data for up to 1990 are sourced from Carbon Dioxide Information Analysis Centre, Environmental Sciences Division, Oak Ridge National Laboratory, Tennessee, United States. Data from 1990 are CAIT data: Climate Watch. 2020. GHG Emissions. Washington, DC: World Resources Institute. https://www.climatewatchdata.org/ghg-emissions. Refer to SP.POP.TOTL for the denominator's source.

Figure 10: Average Price of Electricity in Pacific Countries Compared to World, 2014–2019

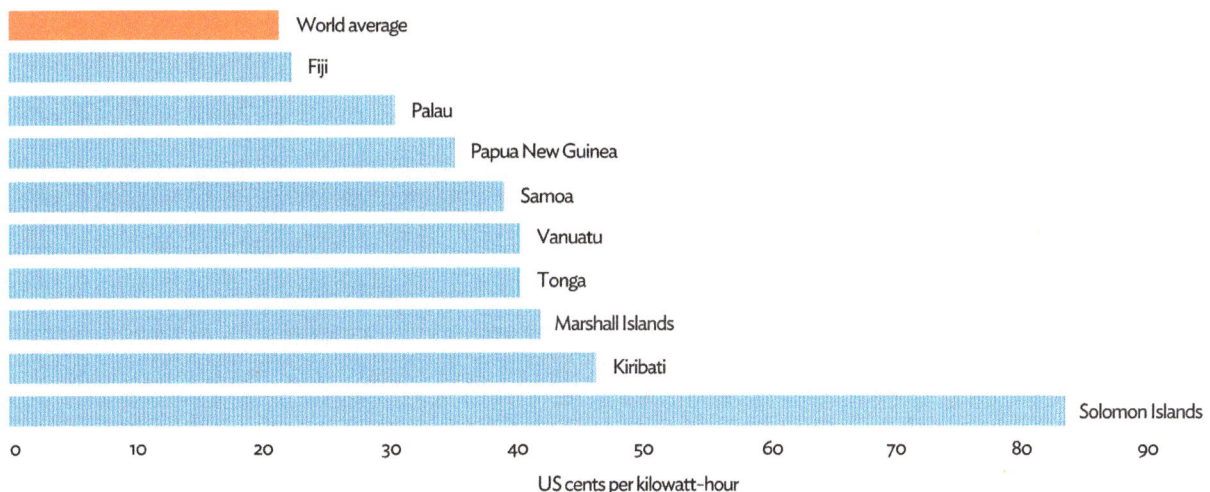

Source: World Bank Group. 2020. *Doing Business Report (DB16-20 Methodology)*. Washington DC.

21   R. Prasad. 2021. The drivers of a clean energy transition in Pacific Island countries (blog). *International Science Council.* 24 September. https://council.science/current/blog/the-drivers-of-a-clean-energy-transition-in-pacific-island-countries/.

To address high electricity costs and improve sustainability, governments in the Pacific and their electric utilities have sought to adopt more renewable sources of power. Renewable energy reduces emissions and generation costs, improves security of supply, and builds resilience to climate change-linked physical and transition risks. The rapid decline in the cost of renewable energy generation and storage systems in recent years has amplified the commercial case for their use.

Between 2010 and 2020, the global weighted-average levelized cost of electricity of newly commissioned utility-scale solar photovoltaic fell 85% (Figure 11). As a result, an estimated 40% of the capacity deployed in 2020 had costs, excluding financial support, that were lower than the cheapest, new, fossil fuel-fired capacity option. Wind power costs have also declined 50% during this period.

While Pacific utilities require smaller-scale generation facilities, they can nevertheless benefit from the rapid fall in solar photovoltaic module prices, which have fallen 93% since 2010.[22]

Despite the advantages of renewable energy, its uptake has been relatively slow in the Pacific (Figure 12)[23] because of a combination of factors, including high transition costs, limited availability of land, and financing constraints. To accelerate the transition, governments in the Pacific have set renewable energy uptake targets, provided financial support mechanisms, and, in some cases, passed laws that require compliance action. International donors are also providing concessional finance and other support to utilities to encourage greater investment in renewable sources of power.[24]

Figure 11: Global Weighted-Average Utility-Scale Levelized Cost of Electricity by Technology, 2010–2020

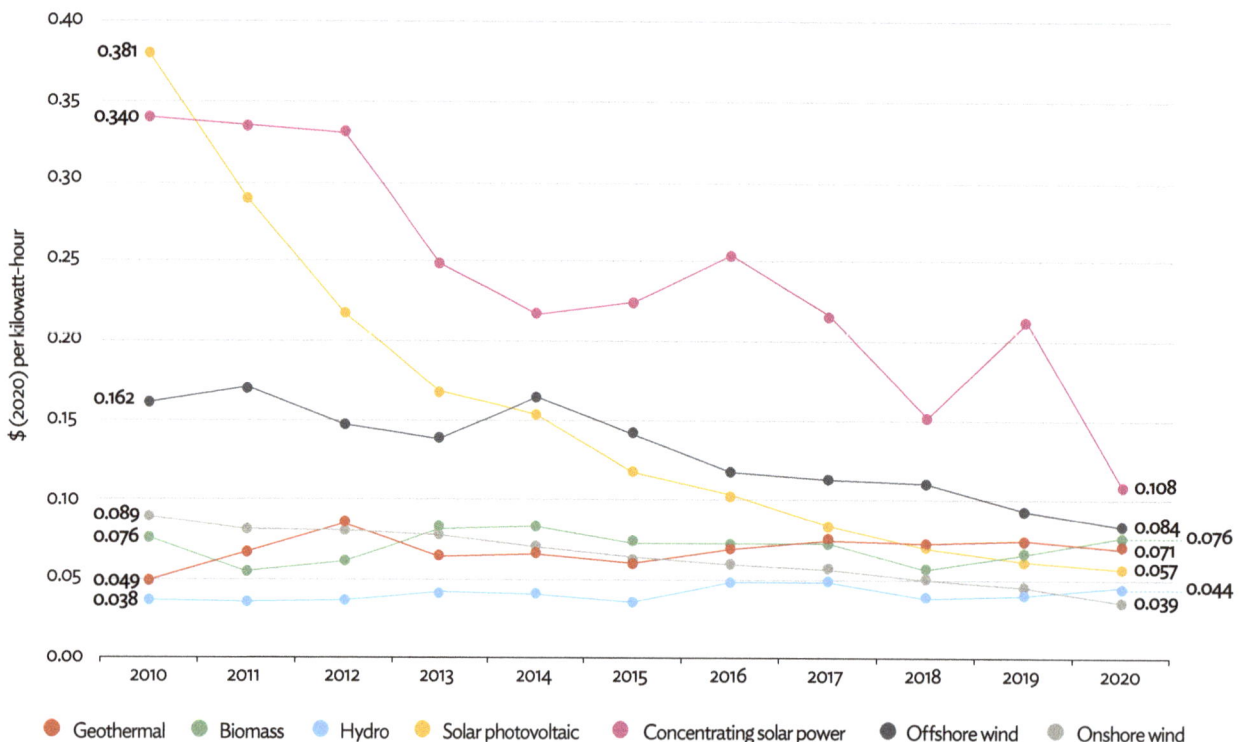

Source: International Renewable Energy Agency (IRENA). https://www.irena.org/publications/2021/Jun/Renewable-Power-Costs-in-2020

[22]   International Renewable Energy Agency (IRENA). Renewable Cost database. https://www.irena.org/publications/2021/Jun/Renewable-Power-Costs-in-2020.

[23]   This data includes the use of traditional biomass as a source of renewable energy. Although subject to many data quality issues, data excluding traditional uses of wood biomass burning suggests that, in 2017, renewable energy supplied only 12.3% of the total final energy consumption on average across the Pacific region (Secretariat of the Pacific Regional Environment Programme 2019).

[24]   Asian Development Bank. 2020. *Pacific Energy Update 2020*. Manila.

Figure 12: Pacific Country Renewable Energy Generation versus Targets

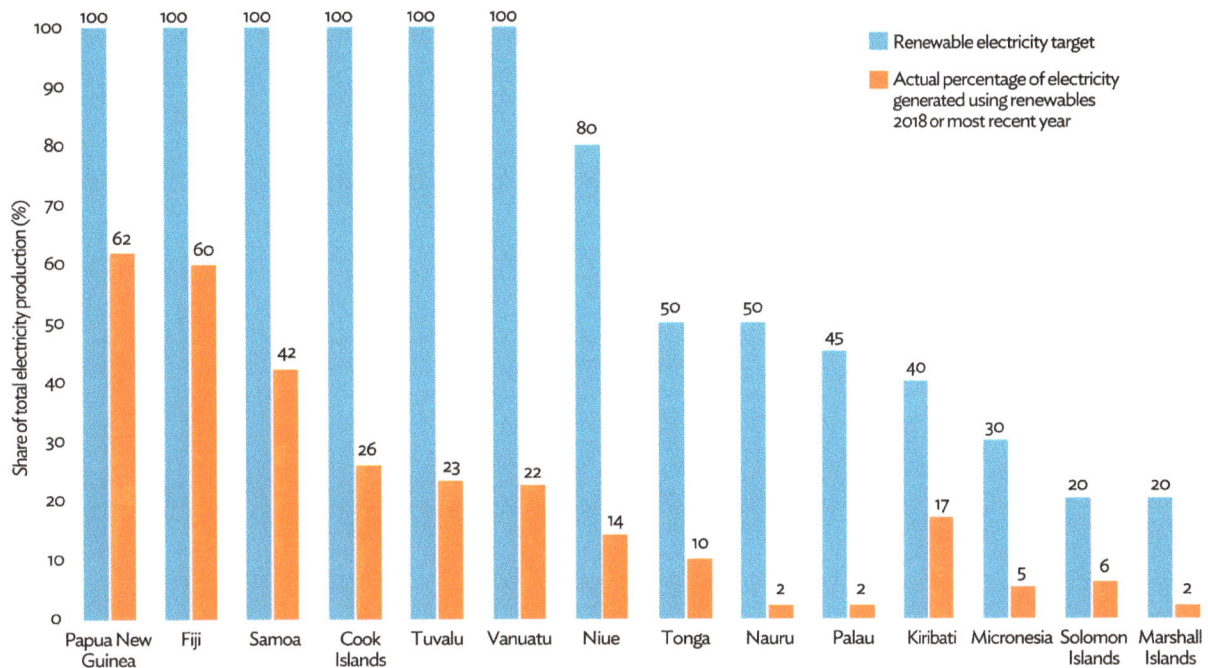

Sources: Secretariat of the Pacific Regional Environment Programme. 2021. *State of Environment and Conservation in the Pacific Islands: 2020 Regional Report.* https://library.sprep.org/sites/default/files/2021-03/SOE-conservation-pacific-regional-report.pdf.

B. Climate Risks Facing Electric Utilities in the Pacific

Although Pacific electric utilities contribute relatively little to global GHG emissions, they face considerable and growing climate-related financial risks. These are generally categorized as "physical", "transition" or "liability" risks:

(i)   **Physical risks** refer to the impacts that climate change may have directly on firms (e.g., company assets damaged by climate change-induced flooding event) or indirectly (e.g., impact of cyclones on supply chains).

(ii) **Transition risks** are those economy- and policy-driven changes which will occur because of climate change. This might include changes in demand for certain goods and services, changes in financing opportunities because of capital flight from emissions-intensive activities, or policy and regulatory changes which may impose compliance costs or other costs on firms (e.g., changes in

demand caused by the transition from fossil fuels used in combustion engines to electric vehicles).

(iii) **Liability risks** relate to firms, their directors, and other representatives being sued because of an action or inaction related to climate change. Currently, liability risks are limited in the Pacific, so are not covered in detail here. However, there is increasing interest in and funding for lawsuits in the global South against high-emitting firms, which could increase the likelihood of this risk for Pacific island utilities.[25]

1. Physical Risks

Electric utilities in the Pacific will continue to encounter significant physical risks associated with climate change, such as coastal inundation, increased frequency and severity of cyclones, and increased temperatures. Recent disasters in the Pacific, such as Cyclone Harold in 2020, have highlighted the significance of these risks. The most recent analysis by the Intergovernmental Panel on Climate Change estimates that, based on current levels of emissions, these climate impacts are likely to intensify and become more frequent as global warming continues.[26]

[25]   J. Setzer and L. Benjamin. 2019. Climate Litigation in the Global South: Constraints and Innovations. *Transnational Environmental Law.* 9 (1). pp. 77–101.

[26]   Intergovernmental Panel on Climate Change. 2022. *Climate Change 2022: Mitigation of Climate Change. Contribution of Working Group III to the Sixth Assessment Report of the Intergovernmental Panel on Climate Change.* Cambridge University Press, Cambridge, UK and New York, NY, USA.

Table 7: Material Physical Risks to the Supply and Demand for Electric Utility Services in the Pacific

Nature of Risk	Description of Risks	Potential Financial Impacts	
Acute	Increased severity of extreme weather events, e.g., cyclones or floods	(i)	Reduced revenue from decreased electricity production. This may arise from disruptions to energy supply chains (e.g., transport difficulties related to input fuels). It could also arise from a reduction in water for hydroelectric facilities (i.e., supply-side shock), or from depressed demand following major weather events (i.e., demand-side shock).
Chronic	Long-term changes in precipitation patterns and other weather patterns	(ii)	Reduced revenue from negative impacts on workforce (e.g., extreme heat distress may lead to greater absenteeism).
	Rising mean temperatures	(iii)	Write-downs of asset values in highly exposed areas (e.g., power plants subject to significant coastal inundation risk).
	Rising sea levels and storm surges	(iv)	Increased operating costs for some assets (e.g., a reduction in precipitation may drive need to purchase water for cooling of some fossil fuel plants).
	Increased stakeholder concern or negative feedback	(v)	Increased capital costs and insurance costs (e.g., power assets damaged from climate events will need to be replaced or repaired, increasing costliness of insurance coverage).

Source: World Business Council for Sustainable Development. 2021. *Evaluating Climate-Related Financial Impacts on Power Utilities.* Geneva.

The consequences of continued global warming may impact both the supply of, and demand for, electricity services. On the supply side, physical climate impacts could damage generation, distribution, or transmission infrastructure, or cause disruption to energy supply chains. Many power utilities have reported these risks already. For example, the Palau Public Utilities Corporation estimates its combined losses at almost $3.7 million from three major typhoons over the past 10 years (Typhoon Bopha 2012, Typhoon Haiyan 2013, and Typhoon Surigae 2021). On the demand side, increasing average temperatures may drive demand for cooling services, pushing up demand peaks.

2. Transition Risks

Transition risks are emerging as the global economy continues to understand and account for the impacts of climate change. Market structures, technology, policy, and regulatory settings are evolving. For example, consumers may prefer "green" electricity production options, governments may create incentives to produce such green power, and financiers may increasingly refuse to finance polluting activities. Some of these social, policy, and economic changes will occur outside of the Pacific, but will

impact the region nonetheless. For instance, major public sector financiers have come under increasing pressure to reduce and/or stop financing fossil fuel projects, which will have an impact on the energy financing available in PacDMCs. Climate change transition risks also arise because of domestic market changes, including in relation to the utilization of existing power assets. For instance, electric utilities are facing commercial pressures as renewable energy prices fall, making distributed energy systems—such as solar and battery—more attractive to customers. The Cook Islands has already implemented a grid-connected household solar program, and Palau is following suit.[27] Although regulatory challenges in some countries limit the uptake of distributed energy technologies, these market changes pose longer-term risks for Pacific utilities.

All PacDMCs have adopted policies that support increasing renewable energy use. Underpinning the commitments made within their Nationally Determined Contributions (NDC) under the global United Nations climate treaty (the 2015 Paris Agreement), PacDMCs are adopting laws to further incentivize decarbonization. For example, the recently passed Climate Change Law in

---

[27]   https://www.adb.org/projects/54011-001/main.

Fiji commits the country to net zero emissions by 2050, meaning EFL will likely come under pressure to further decarbonize. As a consequence of these and other changes, fossil fuel-polluting assets could face significant write-down risks in coming years.

3. Risk Mitigation Activities

Electric utilities can take several steps to manage the physical and transition risks associated with climate change. These include:

(i)   Enabling and investing in flexible low-carbon power sources, such as solar and wind, as well as infrastructure which will smooth out intermittency challenges. This may involve investing in large-scale storage or utility-scale low-carbon facilities, such as hydro.

(ii)   Investing in electricity grid infrastructure to boost resilience to extreme weather, and deploying new technologies to forecast and prepare for extreme weather events.

(iii) Making climate change resilience a central part of utility decision-making, and electricity system policymaking and planning. SOE boards and management must have the skills and tools to forecast and manage relevant climate-related risks.

Table 8: Material Transition Risks to the Supply and Demand for Electric Utility Services

Nature of Risk	Description of Risks	Potential Financial Impacts
Policy and legal	Pricing of greenhouse gas emissions	(i) Increased operating costs arise from cost of compliance with carbon pricing policies and other regulatory requirements.
	Enhanced emissions reporting obligations or other forms of "enforced self-regulation" (see below)	(ii) Write-offs, asset impairment, and early retirement of assets because of legal and policy changes may impact commercial viability of such assets.
	Emission reduction requirements or regulations placed on existing goods and services	(iii) Demand for power reduced because of broader policy changes. For example, incentives for rooftop solar may reduce demand from the grid.
Technology	Substitution of existing power generation, distribution, or transmission technologies with new innovations	(i) Write-offs and early retirement of existing power assets as new technologies come online undermine financials of incumbents. For example, as costs of batteries and solar fall, it may drive down demand for grid-driven power, and this may render the operation of large assets uneconomical.
	Costs for transitioning to new technology or for failing to transition to new technologies	(ii) Capital costs in adopting new technologies, including the technical expertise required.
Market	Change in customer preferences and behaviors	(i) Reduced demand arising from changes in customer preferences for distributed clean energy, rather than fossil fuel-powered energy.
		(ii) Cost of production increased because of increasing input prices, such as for water or taxed fossil fuels.
	Increased stakeholder concern or negative feedback	(iii) Costs associated with shareholder demands for significant change. For example, shareholders may demand significant change in a disorderly way, which may be more costly.

Source: World Business Council for Sustainable Development. 2021. *Evaluating Climate-Related Financial Impacts on Power Utilities.* Geneva.

Some Pacific electric utilities have been taking steps to manage climate-related risks. For example, Tonga Power Limited carried out the Tonga Village Network Upgrade Project, which improved the resilience of existing infrastructure and provided training to technicians. The benefits of these investments are clear. When Cyclone Gita hit the country in 2018, only 10% of the electricity distribution infrastructure in areas covered by the Tonga Village Network Upgrade Project was damaged, compared to 80% in regions outside the coverage of the upgrade works. The Palau Public Utilities Corporation has also developed electricity grid resilience plans, including the installation of substations to improve grid stability in the face of disasters triggered by natural hazards. Other utilities, such as the Solomon Islands Electricity Authority, are developing resilience strategies. In the next section, we consider some of the opportunities and challenges for making these and other types of investments in Pacific electric utilities.

C. State-Owned Electric Utilities and Climate Change Responses

Global research on electricity systems has shown that there are material differences in the way such systems operate, depending on their regulation and level of state ownership. A nascent body of research is considering how state ownership impacts the ability of electricity systems to respond to climate change.

SOEs dominate the supply and distribution of electricity in PacDMCs. With the exception of Union Electrique du Vanuatu (UNELCO), which is not owned by the Government of Vanuatu, and EFL, which is partially privatized, all remaining Pacific electric utilities are wholly state-owned. In this section, we consider how the governance of state-owned utilities might impact their response to climate change.

Recent scholarship among economists and energy policy scholars has focused on the positive role that state ownership might play in responding to climate change.[28] A literature has emerged which associates state ownership with greater levels of investment in clean energy technologies, supported by preferential access to state capital resources.[29] Evidence also shows[30] that, under certain circumstances, state-owned utilities in the European Union are more likely than private firms to invest in renewable energy. Also, evidence suggests[31] that, in some of the largest economies, greater levels of state equity ownership correlate to greater levels of adoption of renewable energy. In these studies, the authors speculate that government ownership gives utilities space to invest in technologies that will yield longer-term financial returns—such as clean energy— because they are not subject to short-term shareholder return expectations.

On the other hand, a well-established literature in economics highlights how state ownership is often associated with chronic, structural inefficiencies. This literature suggests SOEs suffer from a distortion of market incentives by political or other nonfinancial goals, and poor governance.[32] The Finding Balance series has highlighted how such problems have undermined the efficiency of SOEs in the Pacific for many years. Scholars who have studied SOEs and climate change suggest that, under conditions of low administrative capacity, high levels of corruption, or conflicting government policy mandates, it may be harder for such firms to pursue new technologies or processes, including clean energy.[33] How should we make sense of these conflicting perspectives about SOEs and their ability to respond to climate change?

Recent research on SOEs in the electricity sector suggests state ownership alone is not the most important variable for shaping how utilities respond to climate change. Instead, a combination of corporate governance structures and government policy settings shape the willingness and ability of SOEs to reduce climate risks and decarbonize.[34]

[28] M. Mazzucato and G. Semieniuk. 2017. Public financing of innovation: New questions. *Oxford Review of Economic Policy. 33 (1).* pp. 24–48; and M. Mazzucato and G. Semieniuk. 2018. Financing renewable energy: Who is financing what and why it matters. *Technological Forecasting and Social Change.* 127. pp. 8–22.

[29] A. Sterlacchini. 2012. Energy R&D in private and state-owned utilities: An analysis of the major world electric companies. *Energy Policy.* 41. pp. 494–506.

[30] B. Steffen, V.J Karplus, and T.S. Schmidt. 2020. State Ownership and Technology Adoption: The Case of Electric Utilities and Renewable Energy. *MIT CEEPR Working Paper Series No. 16.* Cambridge, Massachusetts.

[31] A. Prag, D. Röttgers, and I. Scherrer. 2018. State-Owned Enterprises and the Low-Carbon Transition. *OECD Environment Working Papers, No. 129.* Paris. https://doi.org/10.1787/06ff826b-en.

[32] B. Holmstrom and P. Milgrom. 1994. The Firm as an Incentive System. *The American Economic Review. 84 (4).* pp. 972–991; A. Shleifer. 1998. State versus Private Ownership. *Journal of Economic Perspectives. 12 (4).* pp. 133–150; and A. Shleifer and R.W. Vishny. 1997. A survey of corporate governance. *Journal of Finance. 52 (2).* pp. 737–783.

[33] F. Belloc. 2014. Innovation in State-Owned Enterprises: Reconsidering the Conventional Wisdom. *Journal of Economic Issues. 48 (3).* pp. 821–848.

[34] A.C. Dibley. 2023. Why does Leviathan innovate? The law and economics of technological change at state-owned electric utilities. *Harvard Environmental Law Review.* Forthcoming.

This analysis suggests the central issue is not the degree of state ownership, but how government policy and corporate governance interact to enable or limit technology switching and/or investment into resilience.[35] We draw below on this latter analysis, which considers the variables beyond state ownership, to evaluate Pacific electric utilities and their decarbonization processes.

D. Interaction of Government Policy and Corporate Governance

1. International Experience

Recent research highlights the interaction of government policy priorities and firm corporate governance in shaping responses to climate change. Government policy refers to the extent to which a state is committed to acting on climate change, both in responding to physical and transition risks and in reducing emissions. Studies have highlighted that SOEs tend to adopt clean energy technologies more quickly when overseen by governments which exhibit stronger concern and policy action on climate change (footnotes 30, 34). Beyond ensuring that sector regulations enable utilities to charge cost-recovery tariffs, governments seeking to respond to climate change can do so through a range of mechanisms, including:

(i)   **Direct targeted mechanisms**—such as mandates for state firms to take adaptation action. For example, under Fiji's Climate Change Act 2021, the minister responsible for climate change can issue guidelines to SOEs to ensure that investment decisions are consistent with broader mitigation targets as set out under the act, and to ensure that climate risks are managed effectively.

(ii) **Enforced self-regulation**—allowing corporations to regulate their own conduct with respect to climate risk management, but insisting self-regulation occurs. These mechanisms can help to build the internal expertise (sometimes referred to as "management-based regulation") to manage climate change.

(iii) **Indirect targeted mechanisms**—such as preferential financing for renewable energy. This might include offering concessional financing, loans, guarantees, or other support. These approaches are widely used in PacDMCs. For example, sections 87–91 of the Fiji Climate Change Act 2021 introduce a framework for the provision of concessional financing from the government and international donors.

(iv) **Economy-wide mechanisms**—such as carbon pricing and other renewable energy incentive mechanisms which encourage all actors in a sector to decarbonize. As highlighted in Table 9, a few PacDMCs have introduced incentive mechanisms, while all PacDMCs have introduced high-level goals for renewable energy growth.

---

[35]   The existing literature is focused on adoption of new energy technologies and not on response to climate risks. However, we draw on this literature here because it focuses on "innovation"; that is, the ability of state-owned enterprises to change processes and technologies. As such, we think the core principles of this literature are applicable to both issues of clean energy adoption and climate risk.

Box 3: Enforced Self-Regulation

Enforced self-regulation is the strategy of allowing enterprises to regulate their own conduct, but insisting self-regulation occurs.[a] The concept of enforced self-regulation is akin to that of management-based regulation. As the name implies, management-based regulation requires regulated enterprises to engage in planning, analysis, and other management activities to address the problem subject to regulation.[b]

Enforced self-regulation and management-based regulation are alternatives to command-and-control regulation or nonregulation. Command-and-control regulatory policies are often unsuitable for addressing societal problems. The same can be said for laissez-faire policies, under which problems are left to be resolved (or not resolved) by market or other forces.

The 2015 Paris Agreement on climate change deploys enforced self-regulation and management-based regulation at an international level. Under the Paris Agreement, countries submit their own mitigation plans, agree to follow certain transparency guidelines for reducing greenhouse gas emissions, and commit to reviewing progress at 5-year intervals.

The Task Force on Climate-Related Financial Disclosures is another important international initiative. The risk reporting framework of the Task Force on Climate-Related Financial Disclosures seeks to achieve consistent climate-related financial risk disclosures for use by companies, banks, and investors in providing information to stakeholders. The framework is widely used in relation to corporate climate risk management, and could be adapted for Pacific electric utilities.

Enforced self-regulation and management-based regulation are used, to some extent, with state-owned enterprises. In Sweden, state-owned enterprises are required to publish annual sustainability reports in which they report on progress across a range of sustainability activities, including climate change. In Papua New Guinea, the Climate Change Act calls on entities to develop plans for emissions reduction and climate change adaptation.

a   This approach is set out in I. Ayres and J. Braithwaite's influential work, Responsive Regulation: Transcending the Deregulation Debate (1992).
b   C. Coglianese and S. Starobin. 2020. Management-Based Regulation. In M. Faure, ed. *Elgar Encyclopedia of Environmental Law VIII*, Chapter 20, pp. 292–307.
Source: Pacific Private Sector Development Initiative.

Table 9: Renewable Energy Targets and Policy Incentives in Pacific Island Countries

Country	Regulatory Policies						Public Financing and Fiscal Incentives			
	RE Target	RE in NDC	Feed-in-tariff	Utility quota obligation / RPS	Net metering	Tendering	Reductions in sales, energy, CO_2, or other taxes	Investment or production tax credits	Energy production payment	Public investment, loans, grants, capital subsidies, or rebates
Fiji	E, P	Y					Y	Y		Y
Kiribati	E, P	Y								Y
Papua New Guinea	E, P	Y								Y
Solomon Islands	E, P	Y								Y
Vanuatu	E, P	Y	Y				Y			Y
Marshall Islands	E, P	Y					Y			Y
Samoa	E	Y								Y
Tonga	P	Y				Y	Y			Y
Palau	E, P	Y	Y		Y					Y
Sri Lanka	P	Y	Y	Y	Y	Y	Y		Y	Y

CO_2 = carbon dioxide, NDC = Nationally Determined Contributions, E = energy, P = power specific target, RE = renewable energy, RPS = renewable portfolio standard,[36]
Y = Yes.
Source: Adapted from the REN21, Global Status Report 2021, Table 6.

The extent to which governments can use the above policy mechanisms to influence SOEs to pursue decarbonization or resilience will depend in part on the corporate governance structures governing the firm. Some of the key variables which influence the responsiveness of state-owned utilities include:

(i) the firm's level of agency—the extent to which the management of the firm can make decisions free of state interference;

(ii) the firm's level of commercialization—the extent to which the firm is empowered to respond to commercial priorities, rather than the political or policy priorities of government;

(iii) the firm's technical capabilities—the administrative and financial capability of the firm; and

(iv) exposure to market competition—the extent to which the government protects the firm through electricity sector policy.

Studies of state-owned utilities have shown that firms which are most responsive to climate change act under different combinations of government policy interest and corporate governance structures. Under conditions where the state has significant influence over the SOE, or the SOE has low levels of agency, low exposure to market competition, and low levels of commercial competence, the role of government policy becomes more significant. In these cases, reducing the cost of renewable energy, or greater market recognition of climate risks, are unlikely to influence utilities. Instead, governments need to proactively direct, finance, or offer incentives to the management of utilities to respond to climate change.

Where SOEs are more independent and commercialized, they are more inclined to respond to commercial opportunities to decarbonize, or to the financial risks associated with climate change. There are examples of state-owned utilities rapidly responding to climate change

---

[36] A renewable portfolio standard is an obligation placed by a government on a utility company, group of companies, or consumers to provide or use a predetermined minimum targeted renewable share of installed capacity, or of electricity or heat generated or sold. A penalty may or may not exist for non-compliance. These policies also are known as "renewable electricity standards", "renewable obligations", and "mandated market shares", depending on the jurisdiction.

under more arm's-length commercial structures. Formerly, Danish state-owned energy company[37] Orsted was an oil and gas company which was heavily influenced by the state, and operated under full state equity ownership. In 2006, the Government of Denmark introduced several commercial reforms, including a merger with five other energy companies, creating a more arm's-length commercial relationship, and reducing the state's shareholding of the merged entity to 73%. This commercial freedom, together with favorable government policies that support renewable energy, enabled the utility to pursue an ambitious strategy in 2009 to decarbonize and become a renewable energy company. In the ensuing 7 years, Orsted raised additional private equity,[38] culminating in its 2016 initial public offering and listing on the Nasdaq Copenhagen, which reduced the state's shareholding to 50.4%.[39] This provided the additional capital and impetus to complete the divestment of its offshore oil and gas infrastructure. It is now regarded as a world leader in offshore wind, and is no longer in the oil and gas business, reducing its climate risk exposure and emissions profile.

2. Pacific Context

Pacific state-owned electric utilities operate in largely similar commercial and government policy environments. All are majority state-owned (except for UNELCO), and subject to heavy state influence and limited competition. Many utilities suffer from capital constraints and—in some cases—technical capacity limitations, particularly in respect to forecasting and managing climate change risks. A notable exception is EFL, the only partially privatized electric utility, with a strategic investor taking up 44% of equity in 2021.

A key distinction in the policy and regulatory environments of Pacific electric utilities is their ability to charge cost-recovery tariffs. While most Pacific electric utilities calculate tariffs based on costs, these are often subject to review and adjustment based on political, rather than commercial, considerations. Where tariffs are set at levels that do not enable full cost recovery, and utilities do not receive an offsetting subsidy from government, it undermines the SOEs' financial sustainability, weakening their ability to meet broader goals, including investing in decarbonization and resilience. In Palau, legislated caps on tariffs have constrained PPUC's ability to recover costs. Reforms undertaken in 2021 have established an

independent regulatory process for setting cost recovery electricity tariffs, which should enhance transparency and reinforce PPUC's commercial mandate, placing it on a more sustainable footing. Where tariffs are set at levels below cost recovery, subsidy payments should be made to the utilities through the CSO framework which is included in the SOE laws of most PacDMCs. This is the case in Fiji, where EFL has a CSO contract with the government under which it supplies electricity to selected rural communities at rates below the cost of delivery. Each year, EFL receives a CSO payment representing the difference between revenues received and the actual cost of delivery.

Most Pacific electric utilities identify cost as a major driver of their renewable energy targets, yet it is the more commercial SOEs that are most proactive in developing a pipeline of renewable energy projects. EFL and Solomon Islands Electricity Authority, two SOEs which operate under a robust commercial framework with an effective cost recovery tariff structure, both have a pipeline of renewable energy projects to be delivered through PPPs in the next 5 years. The planned establishment of an independent tariff regulatory function in Solomon Islands should further support this. PPUC, which has suffered from a weak commercial framework for most of its existence, is undergoing major corporate governance and operational reforms which—together with regulatory reforms—will enable it to function as a commercial enterprise.[40] This coincides with the development in Palau of one of the largest solar PPP contracts in the region. Chronic political interference in the operations of PNG Power, in contrast, has hampered its ability to implement a renewable energy strategy, despite PNG's vast hydro and geothermal potential,[41] and the comparatively high percentage of renewable sources in its energy mix.

E. Policy Recommendations

Given the prevailing conditions in most Pacific electric utilities, it may be difficult for governments to simply mandate them to pursue more ambitious decarbonization agendas or to better account for climate risks. Instead, it may be more useful for governments to support utilities to build up capabilities and capital to invest in climate risk

[37] From 1973 to 2006, Orsted was known as DONG, then as Dong Energy from 2006 to 2016, and finally as Orsted since 2017.
[38] In 2013–2014, an equity placement reduced government ownership to about 65%.
[39] Orsted Annual Report 2016.
[40] Asian Development Bank. 2020. *Report and Recommendation of the President to the Board of Directors: Palau Public Utilities Corporation Reform Program.* Manila.
[41] Bloomberg New Energy Finance estimates that 2% of PNG's 2.5 gigawatts of renewable energy resources are exploited currently. Only about 5% of the country's 4,200 megawatts of technically and economically feasible hydro potential has been developed so far, according to the International Energy Agency.

management. Below, we outline three key priorities on which PacDMC governments could focus.

1. Introduce Mandatory Climate-Related Financial Risk Disclosure Frameworks for Pacific Utilities

The Task Force on Climate-Related Financial Disclosures (TCFD) is used by financial regulators and voluntary industry and investment initiatives as the standard by which companies are assessed for their management of climate-related financial risks. The TCFD encourages firms to develop and disclose the governance structures, firm strategy, risk management, and metrics and targets used to manage climate-related financial risks. Table 10 outlines the main pillars of the TCFD.

SOEs and their shareholders should adopt key elements of the TCFD. While the TCFD sets out a framework for how investor-held firms should assess and manage climate risks, there is no widely-used equivalent to the TCFD for state-owned (and non-listed) companies. As they have a single or dominant state shareholder, SOEs face unique corporate governance and capability challenges. For this reason, enforced self-regulatory or management-based regulation approaches must account for these features. Pacific climate change legislation (Box 4) includes some elements of enforced self-regulation in relation to SOEs, including penalties for noncompliance, but these obligations are piecemeal and should be revised against the TCFD.

Table 10: Summary of Task Force on Climate-Related Financial Disclosure Framework

Governance	Strategy	Risk Management	Metrics and Targets
Disclose the organization's governance around climate-related risks and opportunities	Disclose the actual and potential impacts of climate-related risks and opportunities on the organization's businesses, strategy, and financial planning where such information is material.	Disclose how the organization identifies, assesses, and manages climate-related risks.	Disclose the metrics and targets used to assess and manage relevant climate-related risks and opportunities where such information is material.
Recommended Disclosures	**Recommended Disclosures**	**Recommended Disclosures**	**Recommended Disclosures**
a) Describe the board's oversight of climate-related risks and opportunities.	a) Describe the climate-related risks and opportunities the organization has identified over the short, medium, and long term.	a) Describe the organization's processes for identifying and assessing climate-related risks.	a) Disclose the metrics used by the organization to assess climate-related risks and opportunities in line with its strategy and risk management process.
b) Describe management's role in assessing and managing climate-related risks and opportunities.	b) Describe the impact of climate-related risks and opportunities on the organization's businesses, strategy, and financial planning.	b) Describe the organization's processes for managing climate-related risks.	b) Disclose Scope 1, Scope 2, and, if appropriate, Scope 3 greenhouse gas (GHG) emissions, and the related risks.
	c) Describe the resilience of the organization's strategy, taking into consideration different climate-related scenarios, including a rise in global temperatures of 2.0 Celsius or less.	c) Describe how processes for identifying, assessing, and managing climate-related risks are integrated into the organization's overall risk management.	c) Describe the targets used by the organization to manage climate-related risks and opportunities and performance against targets.

Source: Task Force on Climate-Related Financial Disclosures (TCFD). 2017. *Recommendations of the TCFD (June 2017)*. Basel.

Box 4: Self-Regulation and Management-Based Regulatory Approaches on Climate Change within Pacific Climate Change Legislation

Papua New Guinea: The Climate Change (Management) Act 2015 requires companies to produce emissions mitigation plans (s.65) and climate change adaptation plans (s.74). The act does not outline what the plans should contain, nor how such plans should be benchmarked, but includes penalties for noncompliance. Implementation remains spotty, given the limited capacity of the Climate Change and Development Authority to issue regulations and monitor compliance.

Fiji: The Climate Change Act 2021 provides for enforced self-regulation in relation to disclosure of nonfinancial information in the private sector (Part 15). Section 94 of the act sets out in more detail the risks which investor-owned companies should disclose. With respect to state institutions, including state-owned enterprises, the act only requires companies to be mindful of climate risks when making procurement decisions, which is a much narrower set of standards than those outlined in the Task Force on Climate-Related Financial Disclosure.

Kiribati: The Disaster Risk Management and Climate Change Act 2019 includes a scheme of enforced self-regulation that applies to government ministries, agencies, and bodies (s.12). The action required under s.12 includes "mainstreaming and integration of climate change and disaster risk management considerations in the execution of their regular functions", and having "plans and standard operating procedures in place to protect staff, equipment and infrastructure and to facilitate continued capacity to function following disasters".

Source: Pacific Private Sector Development Initiative.

To strengthen SOE climate risk management in Pacific electric utilities, we recommend the following set of activities, which builds on the legislated frameworks already in place in PNG, Fiji, and Kiribati:

(i) SOE shareholding ministries should encourage state-owned utilities to develop an awareness of, and capability to respond to, climate change-related financial risks. Specifically, this might involve shareholding ministries providing information about climate-related financial risks to SOEs, and issuing questionnaires about key dimensions of the TCFD. The questionnaire could ask about the extent to which SOEs currently account for climate-related risks as part of broader enterprise risk management activities, and whether the SOEs have the capability to evaluate transition and physical risks. Over time, questionnaires might also ask SOEs to report on their climate transition and physical risks.[42]

(ii) SOEs, shareholding ministries, and treasuries should work together to develop capacity to evaluate and manage climate risk exposures. The climate risks which face SOEs are likely to be similar to those faced by sovereigns more generally. In addition, wholly state-owned SOEs will rely on state treasuries to help manage the financial implications of such risks. It would therefore be useful for SOEs to work with economic policymakers to develop analytical tools to understand the risk, and to develop risk mitigation tools.

2. Raise Capital to Invest in Climate-Risk Mitigation

State-owned electric utilities should pursue opportunities to partner with capital providers and technical experts on transitioning to renewable energy and climate resilience. This includes working with development finance institutions, which can combine concessional finance with climate change technical expertise. Concessional financiers may also be able to support utilities to finance the transition to renewable energy. For example, ADB is currently developing an Energy Transition Mechanism, which is being designed to support clean energy transition among state-owned utilities in Southeast Asia.

Financing could also be raised from private capital providers and funds, which are increasingly searching for investment opportunities that reduce emissions. Fiji's 2017 Green Bond raised F$100 million for climate change mitigation and adaptation projects. State-owned utilities could further access blended financing from private and concessional financiers such as the Green Climate Fund. For example, the Energy Development Corporation

---

[42] Over time, state-owned enterprise shareholding ministries could require climate risk management and mitigation to be specifically covered in the electric utilities' business/corporate plan, and progress against mitigation targets reported in their annual reports.

in the Philippines—the country's largest electricity generator—issued a green bond in 2018 which helped the utility raise $90 million for investment in adaptation and resilience activities.[43]

3. Partner with the Private Sector to Accelerate Renewable Energy Uptake

There is a strong business case to pursue renewable energy and better manage the physical and transition risks facing state-owned electric utilities. Pacific electric utilities recognize that clean energy is likely to enable the reduction of tariffs over time, and proactively responding to climate risks will reduce financial costs over the long term. To assist with this transition, electric utilities have been partnering with the private sector. The sale of 44% of EFL to a strategic investor in 2021 has both raised capital and mobilized expertise for further investment in decarbonization and climate resilience. A total of 47 megawatts (MW) of renewable energy generation capacity has been contracted with independent power producers for Fiji, Palau, Solomon Islands, and Tonga. Governments could further facilitate

the use of independent power producers, and improve their VFM, by establishing robust PPP project development and tendering processes and, where necessary, mechanisms to reduce utility payment risks.

This section of the report has highlighted how the current and emerging risks of climate change, and the growing need to decarbonize, impact state-owned electric utilities in the Pacific. We have highlighted how these utilities have taken steps to respond to climate change, but still have work to do to build resilience. Drawing on emerging research in economics and energy policy, we have highlighted how SOE governance and regulatory frameworks might impact the ability of electric utilities to respond to climate change. Given their capital and capacity constraints, we suggest that ongoing efforts to commercialize SOEs will make electric utilities more responsive to climate risks, and facilitate their access to public and private capital providers and technical experts who share these decarbonization objectives. This approach is consistent with ADB's Pacific Approach 2021–2025 and is increasingly evident in the design of its energy sector investments and technical assistance (TA).

---

[43]   World Bank Group. 2021. Enabling Private Investment in Climate Adaptation and Resilience: Current Status, Barriers to Investment and Blueprint for Action. Washington DC.

V. Country Profiles

A. Fiji

1. SOE Portfolio Composition and Contribution to GDP

The Government of Fiji (as of 2021) has majority ownership of 19 active SOEs which are involved in a range of economic sectors, including banking, transport, agriculture, housing, electricity, and postal and financial services. The portfolio is large, with total assets of F$4.2 billion in 2021, 79% of which are held by the four largest SOEs: EFL,[44] Fiji Airways, Airports Fiji Limited (trading as Fiji Airports),[45] and Fiji Development Bank. Eight of the SOEs hold dominant market positions in their respective sectors, and four are regulated as monopolies.[46]

Figure 13: State-Owned Enterprise Portfolio Assets 2021 (F$4.2 billion)

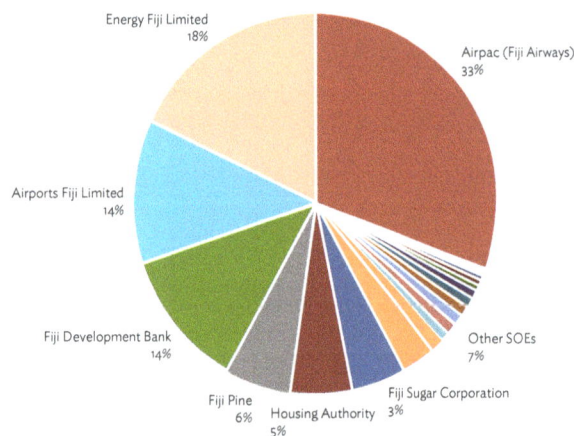

Source: State-Owned Enterprise accounts (audited where available), Ministry of Economy

The portfolio's overall contribution to GDP averaged 3.7% over the 2010–2020 period, substantially lower than the estimated 16%–22% of gross fixed investment in the economy controlled by SOEs in 2020. The contribution to GDP was at a historic low in 2010 because of write-offs at Fiji Sugar Corporation, but improved to an average of 4.1% between 2015 and 2020, despite ongoing losses at Fiji Sugar Corporation and the economic contraction in 2020.

2. SOE Performance 2010–2021

Generally, Fiji's SOEs have been profitable, with an average portfolio ROE of 4% and ROA of 2.1% for the period 2010–2021. Average profitability improved from 2015 to 2021 as compared to 2010–2014. SOEs' ability to efficiently use their assets has declined steadily over the past 10 years, however, from 40% in 2010 to 17% in 2021. In general, the portfolio's growth in assets has not been matched by a commensurate increase in revenue: portfolio assets increased in value by 67% in the 12 years to 2021, while portfolio revenue dropped 31% over the same period. Excluding 2020 and 2021, portfolio assets increased in value by 74% in the 10 years to 2019, while portfolio revenue increased 39% over the same period. This implies that SOEs are, on average, either overinvesting in assets or not charging enough for the services they are providing, or both. Table 11 shows the movement in assets and revenue for the six largest SOEs.

Figure 14: State-Owned Enterprise Portfolio Profitability 2010–2021

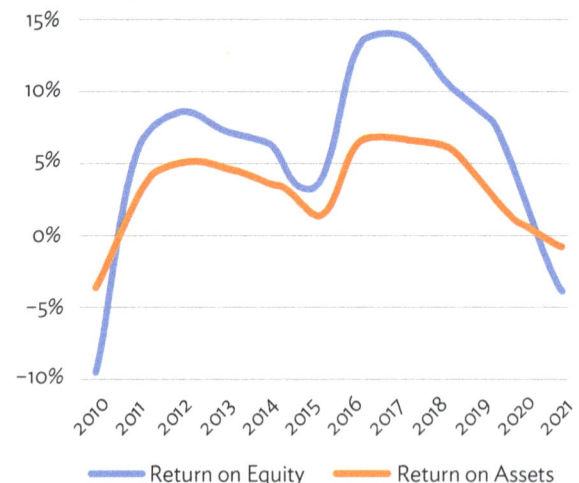

Source: State-Owned Enterprise accounts (audited where available), Ministry of Economy

The portfolio is dominated by the four largest SOEs in terms of assets, two of which contributed 76% of portfolio profits in the period 2015–2021: EFL contributed 44% and Fiji Airports 32%. Fiji Sugar Corporation generated 57% of portfolio losses over the same period. The government

---

44   Previously Fiji Electricity Authority.
45   Airports Fiji Limited commenced trading as Fiji Airports on 29 May 2018.
46   Energy Fiji Limited, Fiji Sugar Corporation, Post Fiji Limited, and Airports Fiji Limited for aeronautical services.

should critically assess whether it retains ownership of smaller SOEs, given monitoring requirements, their lack of strategic value, and their capacity to crowd out private-sector activity.

Table 11: Movement in Asset Value and Revenue for Six Largest State-Owned Enterprises, 2010–2021 (%)

State-Owned Enterprise	Change in Asset Value	Change in Revenue
Fiji Airways	606	(62)
Energy Fiji Limited	(17)	(27)
Fiji Airports	192	(45)
Fiji Development Bank	60	45
Fiji Pine	131	71
Housing Authority	37	(43)

Sources: State-Owned Enterprise accounts (audited where available); PSDI analysis.

3. Preliminary 2021 Results

SOE portfolio[47] financial performance declined substantially in 2021 compared to 2020, as Fiji's economy continued to contract due to COVID-19 border closures. Portfolio revenue declined 17%, and ROE was at -4%, compared to 2% in 2020. ROA, meanwhile, was –2%, down from 1% in 2020.

While Energy Fiji Ltd and Fiji Development Bank managed to maintain the same levels of profitability in 2021 as in 2020, the SOEs exposed to the tourism sector suffered the largest declines. Fiji Airways, which represented the largest proportion (33%) of the portfolio assets in 2021, posted the largest losses, with revenue falling by 37% year on year and ROE falling from –88% to –386%. Airports Fiji Limited, accounting for 14% of the portfolio assets in 2021, also saw a drop in revenue of 44% from 2020, with ROE falling from 0% to –1%. These results are consistent with a 95.4% decline in tourist arrivals to Fiji, from 146,905 to 6,639 between 2020 and 2021. Fiji's coordinated vaccination efforts and planned tourism recovery has resulted in a surge in tourist arrivals from FY2022,[48] which should contribute to improved portfolio performance in 2022.

4. COVID-19 Impact and Response

Several of Fiji's SOEs have been adversely impacted by COVID-19, in particular those involved in international travel. Fiji Airways, Fiji Airports, and Air Terminal Services were impacted by travel restrictions. In the case of Fiji

Airports, many debtors and tenants were unable to make payments when due, and sought payment relief. Fiji Airports also suffered delays in its upgrading of the Nadi terminal facilities and Nausori runway. Fiji Airways took rapid measures to reduce costs, but the collapse in revenues exacerbated its already vulnerable balance sheet. Fiji Airways required government support to increase its debt-to-equity ratio from 4:1 in 2019 to 7:1 in 2020.

Other SOEs impacted include Post Fiji Limited, which saw a significant reduction in letter volume, and Unit Trust of Fiji Management Limited, which had to defer several marketing initiatives and was limited in its ability to service remote customers. Table 12 shows the movement in revenue and net profit after tax (NPAT) for the four SOEs most impacted by COVID-19.

Figure 15: Net Profit 2015–2021 (F$'000s)

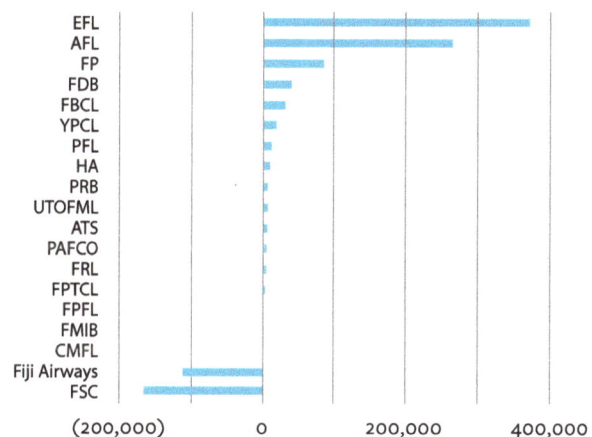

EFL = Energy Fiji Limited, AFL = Airports Fiji Limited, FP = Fiji Pine, FDB = Fiji Development Bank, FBCL = Fiji Broadcasting Corporation Limited, YPCL = Yaqara Pastoral Company Limited, PFL = Post Fiji Limited, HA = Housing Authority, PRB = Public Rental Board, UTOFML = Unit Trust of Fiji (Management) Limited, ATS = Air Terminal Services, PAFCO = Pacific Fishing Company Limited, FRL = Fiji Rice Limited, FPTCL = Fiji Public Trustee Corporation Limited, FPFL = Food Processors (Fiji) Limited, FMIB = Fiji Meat Industry Board, CMFL = Copra Millers of Fiji, FSC = Fiji Sugar Corporation. Source: State-Owned Enterprise accounts (audited where available).

Table 12: Financial Impact of COVID-19

State-Owned Enterprise	Revenue Decline 2019–2021 (%)	Profit Decline 2019–2021 (%)
Fiji Airways	(80)	(678)
Fiji Airports	(81)	(107)
Air Terminal Services	(80)	(149)
Post Fiji Limited	(23)	21

Source: State-Owned Enterprise accounts (audited where available).

---

47   2021 financial statements for all 19 state-owned enterprises surveyed in 2020 were available as of November 2022. Where available, audited accounts were used in the survey.

48   The Fiji Bureau of Statistics has recorded 308,288 arrivals from 1 August 2021 to 31 July 2022.

The decline in NPAT exceeded the decline in revenue for Fiji Airways, Fiji Airports, and Air Terminal Services, which indicates that they were unable to reduce costs at the same rate as revenues declined. The larger the gap between the rate of decline in revenue and the rate of decline in NPAT, the greater the likelihood of financial distress in the short term if revenues do not bounce back to pre–COVID-19 levels. Both Fiji Airways and Fiji Airports were able to restructure their debts and financing facilities, creating breathing room until international borders reopened.

5. Legal, Governance, and Monitoring Framework

SOE ownership monitoring moved in 2019, alongside the enactment of a new Public Enterprise Act (PE Act), from the Ministry of Public Enterprises to a department within the Ministry of Economy. While consolidating SOE monitoring in the Ministry of Economy may have efficiency benefits, it is important that public reporting on SOE performance is enhanced to ensure proper accountability and accessibility (Box 5).

Global practice is moving to specialist bodies responsible for SOE ownership monitoring. A 2018 World Bank[49] study on SOE governance observes that there has been a move to centralize oversight of SOEs through either a dual or centralized model. The dual model is seen as a "transition" model to full centralization, with a single SOE ownership monitoring agency reporting to a minister who is responsible for SOEs.

There are real benefits in undertaking ownership monitoring via an agency separate to the Ministry of Finance, or reporting to a minister other than the minister of finance. It enhances transparency and accountability, as the ownership agency has its own reporting path and requirements separate to the Ministry of Finance, and facilitates contestable streams of advice to cabinet and ministers on the ownership and fiscal impact of the government's investment in SOEs.

The 2019 PE Act improved the governance and commercial framework for SOEs, replacing the 1996 PE Act. Notable improvements include:

- A single minister is responsible for the ownership interest in all SOEs, rather than three ministers under the 1996 act, which led to confused accountabilities and conflicts of interest.

Box 5: Impact of Moving State-Owned Enterprise Ownership Monitoring in New Zealand

In 2009, the Government of New Zealand moved the state-owned enterprise (SOE) ownership monitoring function from the Crown Company Monitoring Advisory Unit, a semi-autonomous unit attached to Treasury for administrative purposes, to a division within Treasury named the Crown Ownership Monitoring Unit (COMU). The ownership monitoring function has been further absorbed within Treasury and is now listed under "Company and Entity Performance Advice" on the Treasury website. Ministerial oversight of SOEs remains split between the two shareholding ministers: the SOE minister with ownership oversight (including primary responsibility for director nominations), and the minister of finance with fiscal oversight.

The demise of the Crown Company Monitoring Advisory Unit, and subsequently COMU, has resulted in a significant reduction in transparency around SOE performance and government interaction with SOEs. Until 2013, COMU produced a comprehensive annual public report on the performance of the SOE portfolio and individual SOEs within the portfolio, including financial and operational benchmarks. While information on performance is still available, it is necessary to search on the Treasury website and the SOE's website to collate the information formerly available through the COMU annual report. The reader must then undertake their own analysis and benchmarking.

Source: Pacific Private Sector Development Initiative.

- Director duties and governance practices are now detailed in the act.
- The definition and legal forms of SOEs are streamlined into a single entity type with the same governance and monitoring arrangements.[50]
- The process for identifying, costing, contracting, and financing community service obligations (CSOs) is clarified.

While the 2019 act strengthened SOE governance arrangements, it only applies to 13 SOEs, representing less than 20% of SOE assets controlled by the government in 2020. The remaining majority-owned SOEs remain outside the Department of Public Enterprises' oversight. Moreover, the 2019 act has some important gaps, notably its reliance on the Finance Management Act 2004 (FM Act) to require SOEs to submit audited accounts. The FM Act is silent on the due date to submit SOE accounts to the responsible minister, and does not require the minister to table the accounts in Parliament, as required in most other jurisdictions. An amendment to the FM Act is planned to address these matters, but has been delayed.

---

[49]   S.C.Y. Wong. 2018. The State of Governance at State-Owned Enterprises. *International Finance Corporation (World Bank Group) Public Sector Opinion. No. 40.* https://documents1.worldbank.org/curated/en/564891520946563480/pdf/NWP-State-of-Governance-PSO-40-PUBLIC.pdf.

[50]   Previously, there were two types of state-owned enterprises (SOEs): government commercial companies (GCCs) and commercial statutory authorities (CSAs). The 1996 act applied different monitoring and reporting criteria to SOEs depending not on their degree of commercial focus, but on how they were incorporated.

6. SOE Reform Highlights 2015–2022

Two privatizations were completed and one PPP contract was signed between 2015 and 2022. The privatizations of Fiji Ports Corporation Limited (FPCL) in 2016 and EFL in 2019 both involved the Fiji National Provident Fund (FNPF). The government retains 41% of FPCL, with FNPF taking up 39% and Aitken Spence Plc 20%. In 2013, Aitken Spence also purchased 51% of FPCL's subsidiary, Port Terminal Services, giving it management control over the Suva and Lautoka ports. In 2019, FNPF acquired a 20% interest in EFL, which it divested to Sevens Pacific Pte Ltd—owned by the Japan Bank for International Cooperation and Chugoku Electric Power Company—in 2021. As part of this transaction, Sevens Pacific also purchased 24% of EFL from the government, bringing its total shareholding to 44%. The government has issued 5% of EFL's shares to domestic customers, and it is intended that those shares will be listed on the South Pacific Stock Exchange (SPX) to enable trading. The government will continue to hold 51% of EFL's issued capital.[51]

In 2021, EFL signed a power purchase agreement with Sunergise Dratabu Pte Limited for the development of a 5MW grid-connected solar plant at Dratabu, Qeleloa. Another PPP, not involving an SOE but notable in its scope, involved the contract to develop and operate the Lautoka and Ba hospitals with a consortium formed by Aspen Medical and FNPF. The 23-year concession contract was signed in 2019.

In addition to the enactment of the PE Act in 2019, other notable reforms include the adoption of Privatization Guidelines in 2017 and a PPP Policy in 2019, the latter replacing legislation that disincentivized the use of PPPs. PPP implementing guidelines have also been developed to support the PPP policy, and were adopted by Cabinet in 2020.

7. Energy Fiji Limited and Climate Change

EFL is a vertically integrated power utility providing electricity to 80% of the Fijian population, while 100% of the population has some access to electricity.[52] Most of the population's energy is supplied through the main grids that run through Viti Levu and Vanua Levu. The grid continues to be expanded through the rural electrification scheme, which is funded by government. Seventy rural electrification projects were completed in 2020. EFL's customer base expanded by 2% to 210,320 from 2020 to 2021. The customer base is made up of 47,525 prepay and 162,795 post-pay.

In 2021, 58.47% of energy generated by EFL was sourced from hydropower, 0.03% from wind, and 6.52% from three IPPs, all involving biomass-sourced generation. The balance of generation is sourced from diesel and heavy fuel oil thermal generators. While the mix in generation capacity between renewable and thermal has been reasonably stable since 2000, the mix in generation source has varied significantly. In 2000, for example, the generation mix between renewable and thermal was about 70:30; in 2006, it was 50:50; in 2007 and 2012, it was 70:30; and between 2013 and 2020, it has fluctuated within a range of 50:50 to 65:35. The mix varies depending on rainfall—to fill the hydro lakes—and fuel prices.

Figure 16: Energy Fiji Limited Fuel Sources 2015 and 2021

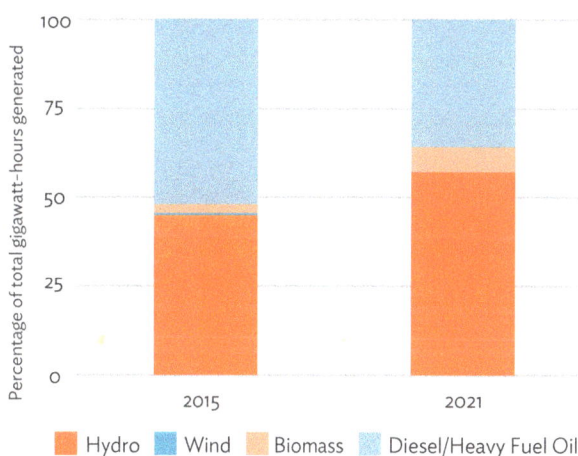

Source: Energy Fiji Limited Annual Reports

While EFL has an ambitious pipeline of renewable energy projects, it has struggled in recent years to successfully tender and negotiate independent power producer (IPP) contracts, with COVID-19 contributing to delays. As a result, EFL has fallen short of the National Development Plan target of 81% of renewable energy generation by 2021, and will struggle to achieve the target of 90% by 2025. A further target of 100% has been set for 2050. The main drivers to increase renewable generation are to reduce generation costs and GHG emissions, and comply with Fiji's Low Emission Development Strategy (2018–2050). Thermal will remain an important back-up generation source to cover years when renewable fuel sources are under pressure because of drought or other factors. The partial

51 It is possible that the government will list its shares on the South Pacific Stock Exchange along with the 5% allocated to customers.
52 https://data.worldbank.org/indicator/EG.ELC.ACCS.ZS.

purchase of EFL by Sevens Pacific in 2021 should help to further prioritize the transition to renewable energy and its commercial benefits.

EFL sustained substantial damage to its power infrastructure in 2020, caused by five tropical cyclones: Sarai in December 2019, Tino in February 2020, Harold in April 2020, Yasa in December 2020, and Ana in January 2021. Total restoration costs incurred by EFL for the four cyclones were estimated at F$4.5 million in 2020, and a further F$7.2 million in 2021. The increasing frequency and severity of tropical storms has heightened the need for EFL to invest in protecting and reinforcing its system assets, which has been an ongoing effort.

EFL's tariff is regulated by the Fiji Competition and Consumer Commission (FCCC); the tariff is reassessed every 3 years, and is designed to provide EFL with a regulated rate of return. The tariff was last reviewed in October 2019 and, as a result, EFL's tariff was increased by 2.74%. The supply of electricity to rural communities is not commercially viable and the government provides funding through a noncommercial obligation (NCO) payment which is reassessed each year.[53] The NCO cost has ranged between F$22 million and F$17 million in the period 2018–2020 and is governed by an NCO agreement between EFL and the government and monitored by FCCC. Where costs exceed the agreed F$22 million cap, the government pays the difference.

COVID-19 had a significant impact on EFL and its operations, with EFL estimating the weighted average reduction in electricity demand between 2019 and 2021 at 12%. EFL also contributed to the government's COVID-19 response. Domestic consumers with a combined household income of F$30,000 or less received a subsidy from government totaling 48% of the cost of the first 100 units of electricity consumed. To mitigate the impact of COVID-19 on households whose incomes had fallen because of lockdowns and loss of jobs, EFL provided an additional subsidy to this consumer group to fund 100% of the cost of the first 100 units, at a cost of F$5 million in 2020. As a result of travel restrictions and lockdowns, EFL's revenue fell 9% in 2020 and 2% in 2021, but the sharp drop in fuel prices and improved hydrology helped the company generate an after-tax profit of F$66.8 million in 2020, and its hedging program helped to offset rising fuel prices in 2021 and secure a profit of F$66.6m.

EFL's financial performance is steadily improving. A sharp drop in profitability in 2014 was largely because of a spike in fuel prices in a year in which low rainfall limited the production of hydroelectricity. Average ROE in the period 2010–2014 was 6%, and increased to 8% in the period 2015–2021. Average ROA for the same periods increased from 3% to 5%. EFL has developed a 10-year Power Development Plan which is updated every 2–3 years, and was last reviewed in 2019. The plan indicates a required investment of F$1.97 billion in generation, transmission, and distribution, and a suite of renewable energy projects. While EFL has determined that bankable projects will be funded through its balance sheet, the company recognizes it will also need to use other financing and delivery options such as PPPs.

Figure 17: Energy Fiji Limited Profitability 2010–2021

Source: Energy Fiji Limited Annual Reports 2010-2021

EFL is entering a new governance and regulatory phase, one in which oversight will be significantly increased and incentives better aligned. The move from a statutory corporation to a company has standardized the governance and reporting framework. There will be three groups now exercising ownership oversight—the government, Sevens Pacific, and customers through the 5% of capital to be listed on SPX. This will create a more robust separation between commercial and political interests, and will strengthen EFL's commercial focus. Combined with FCCC's strengthened regulatory oversight, this move shifts EFL into a governance and regulatory framework found in larger economies, such as New Zealand and Australia, and further away from the Pacific island model of fully government-owned electricity SOEs. PacDMCs should study the developments in Fiji to improve their governance and regulatory frameworks, as

---

many of the reforms undertaken will be applicable, even for smaller economies, or where there is no local stock market.

B. Kiribati

1. SOE Portfolio Composition and Contribution to GDP

Kiribati has 16 active SOEs and is among the largest portfolios in this benchmarking study.[54] The portfolio is involved in a wide range of economic activities including transport, shipping and airlines, banking, utilities, port and airport infrastructure, insurance, and agriculture. The largest five SOEs, Kiribati Ports Authority (KPA), Kiribati Oil Company (KOC), Public Utilities Board (PUB), and Development Bank of Kiribati (DBK), and Airports Kiribati Authority (AKA) accounted for 74% of total portfolio assets in 2020.[55]

SOEs accounted for an estimated 30%–35% of total fixed assets in the economy, and contributed 14.6% to GDP in 2020, about the same as in 2010. SOEs contributed A$0.30 to GDP for every A$1.00 of fixed assets under their control, which is lower than the productivity of the rest of the economy, yet is the highest ratio of the countries in this benchmarking study.

Figure 18: State-Owned Enterprise Portfolio Assets 2020 (A$249 million)

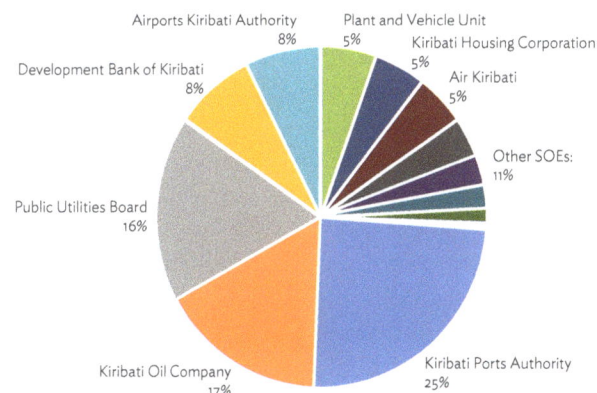

Source: State-Owned Enterprise financial accounts.

2. SOE Performance 2010–2020

While the average ROE and ROA have been reasonably stable over the decade to 2020 at 3.0% and 2.6% respectively, there has been considerable volatility since 2015. ROE fell to 0.5% in 2017 from 5.2% in 2016, while ROA declined to 0.4% from 4.6% over the same period. The main drivers in the decline in portfolio returns between 2016 and 2017 were a 381% increase in fuel costs for Air Kiribati,[56] and Kiribati Ports Authority's recognition of $4.0 million in prior year losses, driving its ROE and ROA from 4.2% in 2016 to –1.4% in 2017. The broad-based SOE reform program that commenced in 2013 following the enactment of the SOE Act appears to have had an immediate impact, lifting financial performance from its 2013 low to its 2016 high. Key drivers influencing the improved performance included strengthened ownership monitoring, enhanced planning and accountability through the development of statements of Intent (SOIs), timely adoption of annual reports, funding-approved CSOs, and transaction reforms such as mergers and privatization or liquidation of non-core or inactive SOEs.

Figure 19: State-Owned Enterprise Portfolio Profitability 2010–2020

Source: State-Owned Enterprise financial accounts.

While implementation of the CSO framework commenced in 2013, most of the identified CSOs were funded from 2017. The major recipients for CSO funding from 2017 to 2020 are Air Kiribati (for flights to and from Kiritimati and a charter agreement with Solomons Air); PUB

---

54   As measured by number of state-owned enterprises (SOEs); only 14 SOEs are counted in the 2020 financial figures presented here, as the others did not have 2020 accounts available at the time of finalization of this study (November 2022).

55   The portfolio financial results presented in this analysis do not include the 2020 accounts of two SOEs which were not available as of November 2022: Central Pacific Producers, Kiribati Oil Co. Ltd, Airports Kiribati Authority, Bwebweriki Net Ltd, and Kiribati Land Transport Authority. All the 2020 accounts made available were unaudited. For all the SOEs that provided unaudited 2020 accounts, except for Kiribati Ports Authority, the 2019 accounts were qualified. Financial results, therefore, should be treated with some caution.

56   Annual fuel costs increased by 87% between 2016 and 2017 and Air Kiribati also paid fuel cost arrears of A$2.5 million in 2017, resulting in a total increase of 381% over 2016.

(for water and sewerage operations); and Kiribati Housing Corporation (KHC) (for rental rates for public servants set below market rates).

Figure 20: Net Profit 2015–2020 (A$ million)

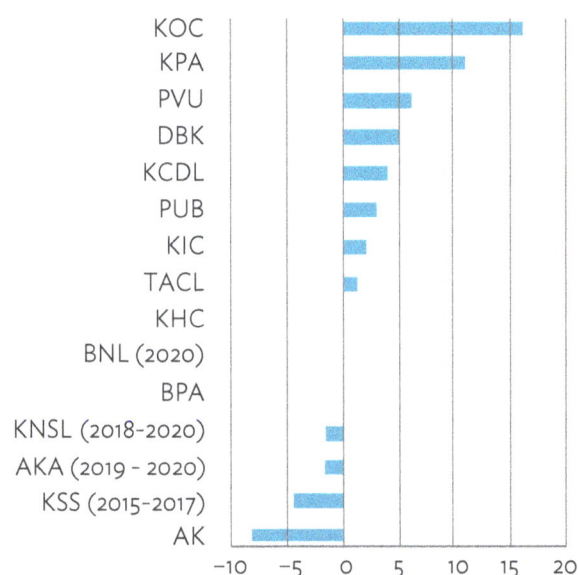

KOC = Kiribati Oil Company, KPA = Kiribati Ports Authority, PVU = Plant and Vehicle Unit, DBK = Development Bank of Kiribati, KCDL = Kiribati Coconut Development Ltd, PUB = Public Utilities Board, KIC = Kiribati Insurance Corporation, TACL = Te Atinimarawa Company Ltd, KHC = Kiribati Housing Corporation, BNL = Bwebweriki Net Limited, BPA = Broadcasting and Publications Authority, KNSL = Kiribati National Shipping Line Ltd, AKA = Airports Kiribati Authority, KSS = Kiribati Shipping Services, AK = Air Kiribati.
Source: State-Owned Enterprise financial accounts.

Two SOEs generated the majority of SOE losses from 2016 to 2020: Air Kiribati and Kiribati National Shipping Limited (KNSL). Kiribati Shipping Services was renamed KNSL in 2018, and continues to be plagued by a lack of maintenance for its ships and the requirement to service noncommercial routes without CSO funding. Air Kiribati has accumulated losses of A$7 million in the period 2015–2020 while receiving CSO funding of A$19.1 million in the period 2018–2020. KNSL[57] accumulated losses of A$5.8 million from 2015 to 2020.

The four most profitable SOEs over the 2015–2020 period were KOC, KPA, Plant and Vehicles Unit, and DBK, which jointly generated 79% of total portfolio profits. The steep increase in profitability between 2018 and 2020, despite the adverse impact of COVID-19 in 2020, was driven by a number of factors. These included the significant CSOs paid to Air Kiribati and KHC, totaling 12.5% of total portfolio revenue in 2020; KPA returning to profit and

achieving an average ROE of 4% in the period 2018–2020, compared to –1% in 2017; reduced losses at KNSL; reduced cost of sales for KOC; and PVU's dramatic increase in revenue. Revenue increased 139% in 2020 over the prior year, while expenses increased 44%. However, as PVU's 2020 accounts are unaudited, and the 2013–2019 audited accounts are qualified, the reported financial position may not be a true reflection of performance.

3. COVID-19 Impact and Response

Kiribati remained COVID-19-free in 2020 because of strict border controls. However, its economy and SOEs have not been immune to COVID-19's impact. The two most impacted SOEs are AKA and Air Kiribati Limited. For the other SOEs, COVID-19's impact has been uneven. Some have reduced operating hours and staff numbers to reduce costs as an offset to declining revenues. International shipping delays have adversely impacted the delivery of machinery and supplies, slowing down some projects, particularly renewable energy projects. The government has provided limited direct financial support to SOEs, usually through existing CSO arrangements. For example, Air Kiribati received a A$7.6 million CSO payment in 2020, allowing it to return a modest profit of A$0.4 million—its first profit since 2017. DBK, with government support, provided some concessional loans to affected SOEs.

4. Legal, Governance, and Monitoring Framework

Kiribati's 2013 SOE Act establishes the legal, governance, and monitoring framework for SOEs. While the act is generally robust, it contains several problematic provisions. The act does not contain a measurable principal objective which serves as an unambiguous commercial mandate. The act requires every SOE to operate as a successful and sustainable business, but defines this as being (i) equally "efficient and effective" as a comparable private business; and (ii) able to service debts without payments from the government, except CSO payments. Measuring comparable efficiency and effectiveness is analytically complex, so the State-Owned Enterprise Monitoring and Advisory Unit (SOEMAU) focuses on profitability metrics, such as returns on equity and assets. The SOE boards, however, are only legally required to meet the principal objective as defined in the SOE Act.

The SOE Act complicates governance by permitting each SOE to have two responsible ministers—the minister of finance and a minister who is assigned by the President to be the minister responsible for the SOE. It is possible for the minister of finance to be appointed the second responsible minister, resulting in one responsible minister. If no minister is assigned, the President acts as the second responsible minister. However, the usual practice is for the minister responsible for the economic sector within which the SOE operates to be the second responsible minister. Sector ministers, however, have a clear conflict of interest in serving as a responsible minister for SOEs in their sector. Their sectoral responsibilities—which include developing sector policy, regulation, and, in some cases, purchasing CSOs—conflict with their ministerial responsibilities to promote the ownership interest.[58] Sector ministers are nominated as either responsible or shareholding ministers in only two other surveyed countries, and this practice is being reviewed in both cases.[59] Kiribati could remove this conflict by appointing the minister of finance as sole responsible minister for all SOEs, as allowed in the SOE Act.

An amendment to the SOE Act, passed in 2016, has further weakened governance. Section 17 of the 2013 SOE Act prohibited the appointment of ministry employees to an SOE board where their employer is accountable to the sector minister. The 2016 amendment allows up to one employee from the ministry accountable to the sector minister to be appointed as SOE director. If appointed, that employee would face a more onerous conflict of interest than that confronting the sector minister serving as a responsible minister—the employee would have to manage the conflict between their role as a ministry employee and SOE director.

The SOE Act encourages skills-based director appointments and prohibits the appointment of members of Parliament. The act also requires biannual director performance reviews. However, the director selection and appointment process has not been codified, and performance reviews are not regularly undertaken. SOEMAU has primary monitoring responsibility for SOEs, and its functions and role are defined in the SOE Act. SOEMAU is established as a unit within the Ministry of Finance, and reports to both responsible ministers. The key monitoring and oversight tools are the SOI, and half-year and annual reports. Both the half-year and annual reports must include

a report on achievement against performance targets in the SOI. The SOI and annual report must be tabled in Parliament. The effectiveness of the SOI as the board's primary accountability document is undermined by enabling the responsible ministers to amend the SOI once adopted by the board.

The SOEMAU's capacity is severely constrained, with each SOEMAU staff member responsible for monitoring six active SOEs. Monitoring staff are also responsible for coordinating the liquidation of SOEs. On average, surveyed countries engaged one monitoring staff member per three to four SOEs. SOEMAU would have been unable to undertake its statutory functions without the support of an ADB TA program that commenced in 2013 and concluded in December 2018. This TA support has also driven the reform program that commenced following the passage of the SOE Act. The Ministry of Finance must now prioritize capacity building within SOEMAU to ensure its effectiveness as an ownership monitor. A new ADB TA program commenced in 2020, with a focus on building SOEMAU capacity.

5. SOE Reform Highlights 2015–2022

Reform commenced in earnest with the passage of the SOE Act in 2013 and the implementation of the act's key accountability and reporting framework—the SOI, half-year and annual reports, and the CSO framework. Ongoing TA is being provided to implement the act and develop monitoring policies and practices within SOEMAU. However, many SOEs continue to struggle to submit reports on time, particularly annual accounts.

Several SOEs have been restructured, merged, or liquidated to reduce losses and improve operating efficiencies. This includes the merger of Kiribati Copra Cooperative Society and Kiribati Copra Mill to form Kiribati Coconut Development Limited, and the liquidation of Bobitin Kiribati Limited, Telecom Services Limited, Betio Shipyards Limited, and Television Kiribati Limited. In the case of KHC, restructuring has enabled it to charge market rents and fund maintenance activities, although ongoing CSO payments are still required.

---

[58]   The conflict created by the sector minister's appointment as a responsible minister is compounded by a 2016 amendment to the State-Owned Enterprise (SOE) Act which allows the appointment of an employee of a ministry, department, or SOE that is accountable to the sector minister to serve on an SOE board. This practice was explicitly prohibited in the 2013 version of SOE Act and results in the sector minister having even greater influence over SOE decision-making.

[59]   Samoa and Solomon Islands.

Box 6: Betio Shipyard Limited

Betio Shipyard was sold through an international tender, with the transaction settled in October 2017. Prior to the sale, the shipyard was in a poor state of repair and could only handle vessels upto 50 tons. Since privatization, the new owner has repaired the slipway, which can now handle vessels of up to 250 tons, and has invested in new equipment including rails, cradle, cables, sand blaster, arc welding equipment, and a refurbished winch. Fifteen local staff have been employed on a permanent basis and another 17 on temporary contracts. Further work has been undertaken to refurbish the office and engineering workshop. A second larger slipway, servicing vessels of up to 1,000 tons, became operational in July 2018. The transfer to private ownership has provided Kiribati access to two working and professionally managed slipways with much greater capacity.

Source: Asian Development Bank. 2021. *Enhancing Economic Competitiveness through State-Owned Enterprise Reform: Technical Assistance Completion Report (TA 8478-KIR)*. Manila.

The government successfully completed a competitive tender for a PPP to rehabilitate the Betio Shipyard in 2017. The shipyard concession has created employment, mobilized private investment, and provided much-needed ship repair facilities to the local shipping fleet (Box 5). The Captain Cook Hotel is also currently being managed under a concession agreement, while the concession for the Otintaai Hotel was terminated in 2017. These transactions have provided the government with valuable experience in structuring PPP transactions.

6. Public Utilities Board and Climate Change

Grid-connected electricity in Kiribati's capital, South Tarawa, is generated and distributed by the PUB, a vertically-integrated water and power utility. The PUB's installed capacity of 7.01MW comprises several diesel generators totaling 5.45MW and a recently completed, grid-connected solar photovoltaic system totaling 1.56MW peak. Outer islands are served largely with solar home systems, and Kiritimati island, the second largest load center (1.65 gigawatt-hours [GWh] in 2016) behind South Tarawa, has a power system not managed by the PUB.[60] PUB provides electricity to 85% of the population on Tarawa, and in 2020 produced 30.3 GWh of electricity, distributing 27.7 GWh.

Renewable energy (solar) represented 7.7% of total generation in 2020, up from 0% in 2015. PUB remains heavily dependent on diesel, which results in high transport costs, price volatility, and insecurity of supply. PUB's transition to renewable energy is primarily driven by cost reduction and supply security aims; however, PUB also recognizes the goals of the government's Integrated Energy Roadmap, Climate Change Policy, and NDC to generate 23% of power from renewables in South Tarawa by 2025.

Studies have concluded that on-grid solar photovoltaic power is the least-cost option for South Tarawa,[61] which has significant technical resource potential for solar energy.[62] Deployment has been limited, however, because of the lack of energy storage to manage intermittency and night time demand, grid instability, weak institutional capacity, affordability concerns, limited financing, and reliance on donor funding.

The South Tarawa Renewable Energy Project, approved by ADB in 2020, will finance the installation of 4MW of solar photovoltaic generation capacity and a 5MW/13 megawatt-hour (MWh) battery storage system. This will enable South Tarawa to increase its renewable energy grid penetration from 9% to more than 44%, exceeding the government target for South Tarawa of 23% contribution by 2025.

Figure 21: Public Utilities Board Profitability 2010–2020

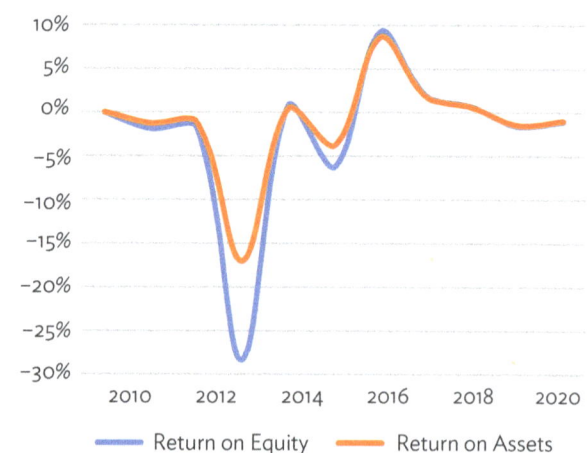

Source: Public Utilities Board financial accounts

PUB's average ROA was 1% from 2015 to 2020, an improvement from –4% from 2010 to 2014, yet insufficient to cover its costs of capital. PUB is responsible for water and wastewater service provision, supported by

---

60   https://www.adb.org/sites/default/files/project-documents/49450/49450-021-pfrr-en.pdf.
61   Levelized cost of $245 per megawatt-hour (MWh) compared with $347 per MWh for diesel.
62   https://www.adb.org/sites/default/files/project-documents/49450/49450-021-pfrr-en.pdf.

CSO payments ranging from A$1 million to A$2 million per year, while the electricity division does not receive CSO support. Electricity revenue represents about 90% of total revenue. Non-core revenue, comprising CSO payments and amortization of donor-funded assets, are significant contributors to total revenue (20%) in 2020. If the amortization of donor assets, which is not a cash item, was removed from revenue, PUB would have incurred losses in 4 of the 5 years since 2015. Nonpayment by customers, including the government, resulted in receivables peaking in 2016 at $9.0 million, or 20% of total assets, while payables—mainly to KOC—also peaked in 2016 at A$7.8 million, representing 60% of total liabilities. In 2020, A$7.9 million of receivables were provisioned as a doubtful debt, while A$7.7 million was owed to KOC for nonpayment of fuel.

COVID-19 has had minimal impact on PUB's operations, other than delaying the implementation of renewable projects because of shipping constraints and closed borders. Although Kiribati is low lying, PUB advises it has suffered minimum impacts from extreme weather events to date. Management advises that poor investment in infrastructure and the absence of preventative maintenance have created a greater immediate challenge to operations than adverse weather events. In February 2018, one of PUB's generators suffered a major fault that took 2 months to repair and cost A$180,000, while intermittent blackouts have occurred since 2020 because of load shedding resulting from generation unit failures. Major repairs were required in 2021 to PUB's three other generators.

C. Marshall Islands

1. SOE Portfolio Composition and Contribution to GDP

The Government of the Marshall Islands has majority ownership of 11 SOEs engaged in a range of economic sectors including transport, banking, telecommunications, utilities, commodities, and transport infrastructure.[63] The five largest SOEs, Republic of Marshall Islands Port Authority (RMIPA), Marshall Islands Development Bank, Marshalls Energy Company (MEC), National Telecommunications Authority, and Air Marshall Islands represent 91% of total portfolio assets, and have contributed 80% of total profits

in the period 2015–2020. SOEs' contribution to GDP has averaged 11% in the 5 years to 2020, and has shown a steady decline. The contribution was 9% in 2020, despite SOEs representing an estimated 20% of the country's total capital stock.

Figure 22: State-Owned Enterprise Portfolio Assets 2020 ($198 million)

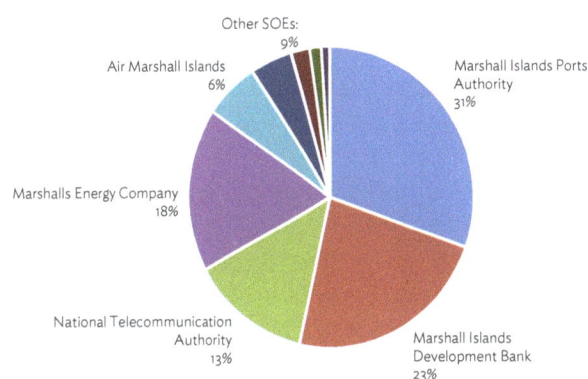

Source: Office of the Auditor General, Republic of Marshall Islands

Many of the country's SOEs are monopolies or dominant in the sectors within which they operate, making their performance, both financial and operational, of significant consequence. Monopoly providers include Kwajalein Atoll Joint Utilities Resources (KAJUR); MEC, the vertically-integrated electricity utility; National Telecommunication Authority; RMIPA, which operates both the port and airports; Majuro Water and Sewer Company; Majuro Atoll Waste Company, which is responsible for refuse collection and disposal; and Tobolar Copra Processing Authority (TCPA), which purchases copra at a government-set price typically above market price. Dominant sector players include the Marshall Islands Development Bank; Majuro Resort Inc. (MRI), the largest hotel in the Marshall Islands; Air Marshall Islands; and Marshall Islands Shipping Corporation, which operates subsidized shipping and ferry services.

Currently, the Marshall Islands does not have anti-trust legislation or an agency which reviews transactions for competition-related concerns.[64] Most SOEs are subject to price control exercised by the ministry or department responsible for the relevant economic sector, subject to Cabinet approval.

---

63 The data presented on the state-owned enterprise portfolio does not include the 2020 figures for Kwajalein Atoll Joint Utilities Resources as these accounts were not available as of November 2022.

64 Government of the United States, Department of State. 2020. *2020 Investment Climate Statements: Marshall Islands.* Washington, DC.

2. SOE Performance 2010–2020

Portfolio ROE and ROA for the period 2010–2020 averaged 1.2% and 1.3% respectively, and improved dramatically after 2015. The average portfolio ROE in the period 2015–2020 was 7.0% and ROA 4.6%. This improvement began in the same year the SOE Act became law. While core parts of the act, such as the establishment of the SOEMU and development of forward-looking business plans, were not implemented until 2018, the increased focus on SOE performance prompted by the SOE Bill's development had positive flow-on effects and contributed to improved results.

The government made significant investments in SOEs in the period 2015–2020. Portfolio ROE averaged 8.3% in this period and SOEs retained all of their profits, adding $48.8 million to shareholder funds. This was mainly applied to debt reduction—debt declined by 51%—and a buildup in cash reserves, which increased by 364%. However, the book value of assets declined by 9% in the period 2015–2020.

With cash now comprising 12% of total assets, it is important that the justification for future investments is articulated in SOE corporate plans. SOEs in all jurisdictions are skilled at hoarding and hiding cash, so governments and their SOEMUs must critically review SOEs' investment plans to ensure that cash is wisely spent. Asset utilization fell from 49% in 2015 to 39% in 2020, indicating that the portfolio assets are being used less efficiently. This could be a sign of aging infrastructure or prices set below true cost.

The government provides significant financial support to SOEs. Table 13 lists the government's contributions in 2020, the most significant being the $8.3 million paid to TCPA to fund the copra price subsidy. Cumulative copra price subsidies paid through TCPA in the period 2015–2020 total $29.7 million. Total government grants and subsidies in FY2020 represent 23% of total SOE portfolio revenue, while the average annual subsidy in the 5 years to 2020 was 22.6% of total portfolio revenue.

Figure 23: State-Owned Enterprise Portfolio Profitability 2010–2020

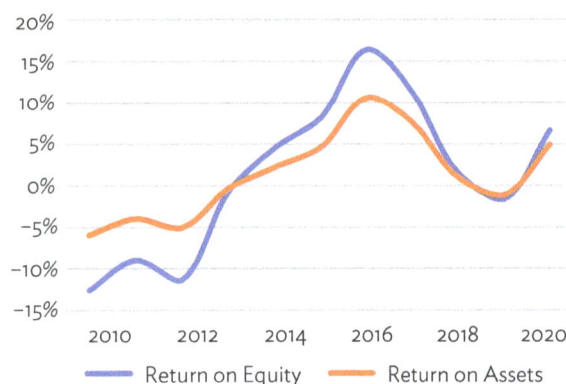

Source: Office of the Auditor General, Republic of Marshall Islands

Table 13: Government Financial Support for State-Owned Enterprises, FY2020[a]

State-Owned Enterprise	Government Grants[b] and Subsidies ($)	Reason and Other Donor Support
MEC	4,032,983	Including $400,000 cost recovery support for Wotje and Jaluit power stations
NTA	1,590,000	Operating subsidy linked to repayment of long-term debt
RMIPA	147,000	Including federal grant of $97,000
MIDB	181,668	Federal grants
MWSC	757,902	Operating subsidies and a further $291,312 from the Government of the United States
TCPA	8,326,938	Annual copra subsidy from the government
AMI	843,000	Operating subsidy
MAWC	598,898	Grant from the government
MISC	2,350,000	Subsidy for shipping services and ferries because of price set below cost. Further $3 million for ship purchases in period 2017–2020

AMI = Air Marshall Islands, MAWC = Majuro Atoll Waste Company, MIDB = Marshall Islands Development Bank, MEC = Marshalls Energy Company, MISC = Marshall Islands Shipping Corporation, MWSC = Majuro Water and Sewer Company, NTA = National Telecommunications Authority, TCPA = Tobacco Copra Processing Authority, RMIPA = Republic of Marshall Islands Port Authority.

a   State-Owned Enterprise audited annual accounts. Excludes any subsidy to Kwajalein Atoll Joint Utilities Resources.

b   In some cases, the government grant is a pass through from an external funder such as $1.9 million to MEC, which is budget support from the European Union passed through government, and $2.1 million to MEC, which is grant aid from Japan.

Source: Ministry of Finance State-Owned Enterprise Monitoring Unit, Office of Auditor General

Despite overall portfolio profitability, five SOEs are chronic loss generators. KAJUR has not produced audited accounts for 2020, but cumulative losses in the period 2015–2019 totaled $2 million, despite government grants of $8.5 million over the same period. RMIPA generated cumulative losses of $4.5 million in the period 2015–2020 while receiving $16.3 million in government grants.[65] In the period 2018–2020, government support for RMIPA declined significantly, resulting in cumulative losses totaling $12.6 million. The Majuro Water and Sewer Company's cumulative losses in the period 2015–2020 were $1.2 million, while the SOE received government support of $2.8 million over the same period. MRI generated losses of $0.21 million in the period 2015–2020 after receiving government support of $0.30 million. Majuro Atoll Waste Company received government support of $5.2 million in the period 2015–2020, yet accumulated losses of $0.07 million.

Figure 24: Net Profit 2015–2020 ($'000s)

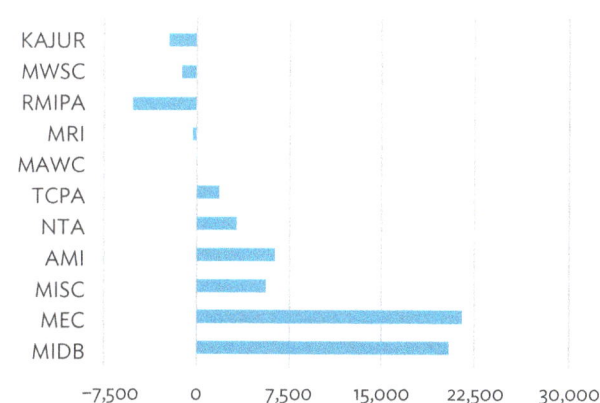

KAJUR = Kwajalein Atoll Joint Utilities Resources, MWSC = Majuro Water and Sewer Company, RMIPA = Marshall Islands Ports Authority, MRI = Majuro Resort, MAWC = Majuro Atoll Waste Company, TCPA = Tobolar Copra Processing Authority, NTA = National Telecommunication Authority, AMI = Air Marshall Islands, MISC = Marshall Islands Shipping Corporation, MEC = Marshalls Energy Company, MIDB = Marshall Islands Development Bank.
Source: Office of the Auditor General, Republic of Marshall Islands

65   Delays in major infrastructure works adversely impacted Republic of Marshall Islands Port Authority revenue in the 2015–2020 period.

These five SOEs require significant structural and operational reforms. Where government support and subsidies are warranted, they should be structured as formal CSOs so there is transparency around cost, funding, outputs, outcomes, and service quality standards. The SOE Act has a robust CSO framework that requires all CSOs to be costed and contracted. The price set for delivering the CSO should include a profit margin, and the service contract should include performance markers. The government has commenced the implementation of the CSO framework, and it anticipates that the first performance-based contract will be signed off in 2022.

3. COVID-19 Impact and Response

Two SOEs have been the most impacted by COVID-19: RMIPA and MRI. RMIPA operates both the port and airport—airport revenues declined 39% in 2020 compared to the year prior because of closed borders and reduced international flights. The seaport division also experienced a 24% drop in revenue because of a decline in shipping vessels visiting the Marshall Islands. The Majuro Resort, the largest hotel in the Marshall Islands, experienced a 36% decline in room rental revenue in 2020 compared with 2019. Food and beverage revenue fell 9% compared to 2019. Total occupancy declined 35% year on year, despite management dropping room rates by 4%. In an attempt to increase local occupancy and income in 2020, the hotel offered special weekend discounts, home delivery of food, and public access to the hotel's laundry services.

In contrast to many Pacific airlines, AMI did not generate operating losses in 2020. The operating revenue of AMI, which operates flights and provides ground handling services at the international airport, increased 2% in 2020 compared to 2019. This was because of an increase in charter revenue for COVID-19 preparedness and repatriation flights, and medivac flights linked to a dengue outbreak in late 2019. This increased flight revenue offset the sharp drop in ground handling revenue. Operating profitability increased 40% in 2020 as fuel costs dropped and AMI reduced general and administrative expenses.

4. Legal, Governance, and Monitoring Framework

A comprehensive SOE Act was enacted in 2015 and establishes an unambiguous commercial mandate for all SOEs—they must be profitable and maximize the net worth of the public investment in the SOE. Section 618 of the 2015 Act prohibited the appointment of public officials as SOE directors, except for a limited transitional term of no more than 3 years from the date the act became law, and then only one public servant per board. Public officials are defined as members of Parliament, ministers, senators, or employees of the public service. Public servants that serve in the ministry with policy or operational responsibility for the SOE's principal business cannot be appointed to that SOE's board. The act also prohibited the appointment of a public official as chair.

The act was amended in 2016 to permit the appointment of ministers to SOE boards, and expanded the number of public officials that could be appointed to three per board. The prohibition against appointing public officials as chair was also repealed. The rationale behind the 2016 amendment is not clear because Parliament had recognized in its approval of the 2015 act that the presence of ministers and public servants on SOE boards created conflicts of interest, led to politicized decision-making, and resulted in poor operational and financial outcomes.

The SOEMU resides within the Ministry of Finance and became operational in mid-2018. Aided by donor partner support and capacity-building efforts in 2018 and 2019, the unit has made progress in implementing key provisions of the SOE Act, building awareness and providing training and guidance manuals for the SOEs and SOEMU. However, further support will be required for the SOEMU to provide effective ownership oversight. The unit consists of three staff members, two of which undertake analytical functions. Most Pacific islands SOEMUs have a ratio of one analyst for every three or four SOEs, while Marshall Islands has one staff analyst for every six SOEs. Further, in the early stages of reform and implementation, the workload is usually heavier.

The SOE Act requires all SOEs to prepare and publish annual statements of corporate intent (SCIs), and they must be tabled in Parliament. The act also requires SOEs to prepare and publish their audited annual reports no later than 3 months after the end of the financial year. The annual report must provide readers with an informed assessment on the SOE's performance, and report on actual progress against performance targets contained in the SCI. The audited annual account must be tabled in Parliament within 15 days[66] of its submission to the minister responsible for SOEs, and a summary published in local newspapers.

---

[66] Section 634(5) of the State-Owned Enterprise Act requires the minister to table the annual report within 15 sitting days of Parliament from the date the minister receives the report.

The SCI and audited annual accounts are key SOE accountability documents, and failure to prepare, adopt, and publish them in a timely manner undermines the transparency and accountability framework established by the SOE Act. Actual practice falls well short of the act's requirements. No SOE develops an effective SCI, and none are published on SOE websites, the Ministry of Finance website, or Parliament's website. Annual audited accounts do not contain reports on progress against SCI targets, summaries are not published in local newspapers as required by the act, and no accounts are found on Parliament's website. Audited annual accounts are published on the auditor general's website, but usually not until after the due date. The audited reports for FY2021, for example, should have been published by 31 December 2021, but only three of the 11 SOEs had by November 2022 published their 2021 audited accounts. Their usefulness as an accountability document has thus been significantly eroded. No SOE has published recent audited annual accounts on their website.

5. SOE Reform Highlights 2015–2022

The major SOE reform in the 2015–2022 period was the adoption of the SOE Act in 2015. The 2016 amendment removed an important tenet of the act by permitting ministers and a larger number of public officials to serve on SOE boards. In 2018, there were significant efforts to build capacity within the SOEMU, which included developing monitoring manuals, monitoring procedures, and documents to support the implementation of the CSO framework established in the SOE Act. However, a lack of capacity within the SOEMU has limited the unit's ability to perform an effective monitoring role. The presence of ministers and other elected officials on SOE boards further undermines the unit's ability to monitor performance and hold boards to account, particularly when the minister is also the board chair. Nevertheless, progress has been made—nine SOEs are now submitting formal planning documents, including financial forecasts, and identifying and costing CSOs. In 2021, ADB approved a $2 million grant to support, amongst other initiatives, continued SOE reform. Funding will be applied to implement the CSO framework and to support gender diversity by providing SOE director training for women candidates.

Sustained improvement in the financial and operational performance of SOEs will require adherence to the SOE Act, adequate resourcing of the SOEMU, and a commitment to depoliticized decision-making.

Removing public officials, particularly ministers and senators, from SOE boards will help SOEs focus on commercial outcomes and improved performance.

6. Marshalls Energy Company and Climate Change

MEC is the vertically integrated monopoly supplier of electricity within the Marshall Islands. It is also responsible for all fuel imports. The company generates, distributes, and retails electricity on Majuro to a population of about 28,000, or about half of the population of the Marshall Islands. The Majuro system, which includes the capital, accounts for 72% of electricity generated and consumed. Ebeye and Kwajalein atolls account for an additional 24%, and the outer islands make up the balance. Power is provided through solar home systems in the outer islands. The power network in Majuro comprises three distribution feeders connected to a power generation station. The level of power interruptions is one of the highest in the Pacific.[67] Power sector losses make up 22% of total energy generation, and are attributable in part to suboptimal equipment specifications and the poor condition of the distribution grid. The grid is susceptible to natural hazards, with an average elevation of less than two meters above sea level.

Renewable power generation is negligible: all power generated in Majuro uses diesel fuel. The government has adopted ambitious GHG emissions reduction and renewable energy targets based on its contributions under the Paris Agreement. A roadmap has been developed for a 65% reduction in diesel consumption for power generation by 2025, and 100% by 2050. Work will also need to be undertaken to improve the distribution grid to address current issues, and to accommodate the large-scale addition of intermittent renewable energy technologies. The grid investment will be staged through to 2030, and is expected to cost $170 million.

The Marshall Islands is highly exposed to natural hazards, including tropical storms and associated tidal surges, coastal flooding, extreme precipitation, and drought. The archipelago of 29 low-elevation coral atolls and five islands has an average elevation of less than 2 meters above sea level, and is vulnerable to natural hazards and climate change impacts. A study by the Pacific Catastrophe Risk Assessment and Financing Initiative indicates the country could incur an average annual loss of $3 million because of natural hazards, while the low-lying Majuro atoll—the

---

67   Asian Development Bank. 2017. *Regional Capacity Building and Sector Reform for Renewable Energy Investment in the Pacific.* Manila. See also: Pacific Power Association. 2018. *Pacific Power Benchmarking Report.* Suva

main population center—is extremely vulnerable. The MEC electricity network is not reliable and is susceptible to natural hazards that damage electrical infrastructure. The Marshall Islands identifies the need for infrastructure resilience in its 2020–2030 National Strategic Plan.[68]

MEC, with its development partners, will invest in climate resilience over the 2022–2025 period in preparation for the introduction and scaling up of renewables. This will include the rehabilitation of the Majuro tank farm, installation of protection devices, and increasing MEC's capacity to plan for and mitigate disaster risks. Currently, MEC neither quantifies nor discloses climate change-related risks or mitigation measures in its annual report.

MEC's profitability has been volatile over the 2010–2020 period, but the SOE has averaged a 9% ROA. Profitability surged in 2020, mainly because of reduced fuel prices. Cumulative profits were $21.5 million in the period 2015–2020, making MEC the most profitable SOE in the portfolio over that period. Subsidies and grants totaled $13.4 million over the same period. Cash reserves have increased from just under $1 million in 2015 to more than $11 million in 2020. While MEC will require significant investment to achieve the country's climate change and service improvement targets, its level of profitability and buildup in cash reserves indicate a review of subsidy contributions and/or tariff structures may be appropriate.

Figure 25: Marshalls Energy Company Return on Assets 2010–2020

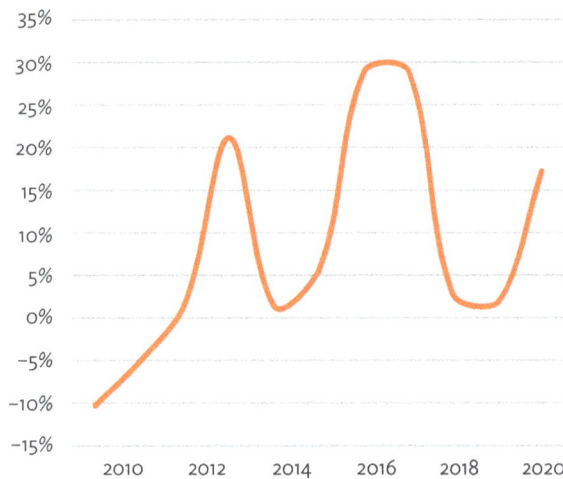

Source: Office of the Auditor General, Republic of Marshall Islands

D. Palau

1. SOE Portfolio Composition and Contribution to GDP

Palau's SOE portfolio is composed of four entities which are the dominant providers of power, water, telecommunications, and banking services. The newest SOE, Belau Submarine Cable Corporation, was established in September 2015 to own and operate Palau's only communications cable, and filed its first financial statements in 2016. The portfolio is large, with a book value of assets of $198 million in 2020, representing about 19%–23% of total fixed assets in the economy. However, the portfolio contributes relatively little to GDP, averaging 4% per year between 2015 and 2020. In 2020, SOEs contributed an estimated $0.13 to GDP for every $1.00 in fixed asset investment, compared to an estimated contribution of $0.60 for the rest of the economy.

Figure 26: State-Owned Enterprise Portfolio Assets 2020 ($198 million)

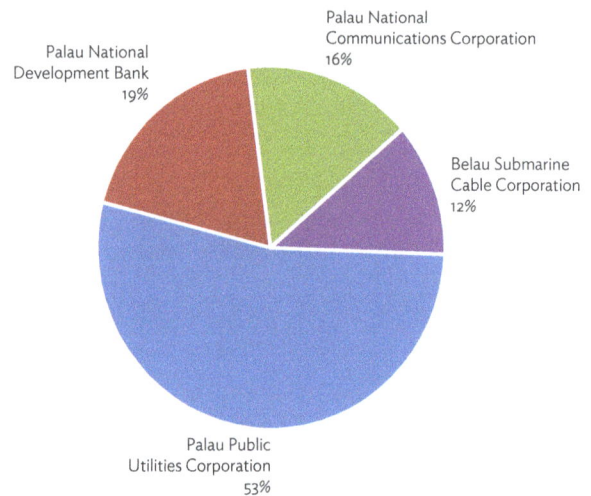

Source: State-Owned Enterprise audited financial statements

About 84% of Palau's public debt was owed by SOEs in 2020, and the Palau Public Utilities Corporation (PPUC) was the largest borrower, with outstanding debt of $23 million (44% of SOEs' total debt, or 9% of GDP). The financial health of SOEs has a direct impact on macroeconomic stability and the government's ability to invest in the infrastructure necessary for economic growth.

68   https://www.adb.org/sites/default/files/project-documents/49450/49450-026-ifr-en.pdf.

2. SOE Performance 2010–2020

Portfolio ROE and ROA averaged –3% and –1% for the 2015–2020 period, among the lowest in this benchmarking sample. Portfolio profitability is largely driven by the results of the Palau National Communications Corporation (PNCC) and PPUC, which generated a net gain of $10 million and loss of $18 million respectively, over the 2015–2020 period. A sharp rise in overall profitability in 2016 is attributable to increases in core revenue for PNCC and operating subsidies and grants for PPUC, while operating expenses were maintained or reduced. In the 2015–2020 period, PPUC turned a net profit in 1 year (2016). In 2019 and 2020, PPUC's core revenue remained flat while PNCC experienced a 15% and 10% drop respectively, as it struggled to grow GSM and digital TV revenue in the face of increased competition from internet-based platforms. Both companies generated losses in 2019 and 2020.

Figure 27: State-Owned Enterprise Portfolio Profitability 2015–2020

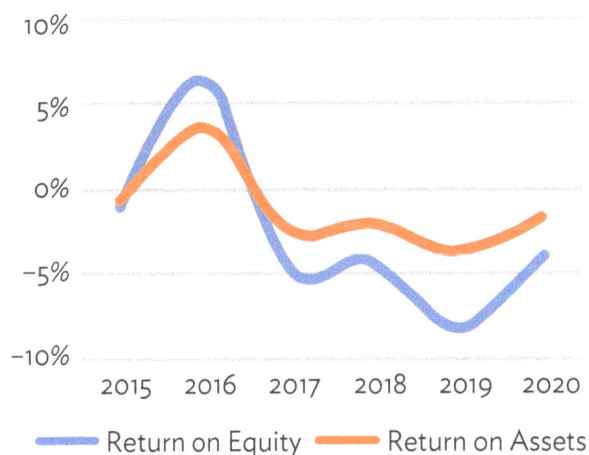

Source: State-Owned Enterprise audited financial statements

Palau's SOEs suffer from a number of structural weaknesses, most notably the absence of a clear commercial focus. This, and the lack of an ownership monitoring framework, has complicated government and board attempts to set consistent performance expectations, and enable the SOEs to achieve them. Palau is one of only two countries in this benchmarking sample that does not have an SOE law to govern the relationship between the legislature, the minister of finance as shareholder, the SOE board, and SOE management. In the absence of such guidelines, the roles and responsibilities of these actors may or may not be specified in the establishing legislation of each SOE, resulting in a lack of coherence and accountability.

3. COVID-19 Impact and Response

Palau's economy, which is heavily reliant on tourism, contracted 10.3% in 2020 as visitor arrivals dropped 81% compared to 2019. The government's response, through the $20 million CROSS Act and the United States Coronavirus Aid, Relief, and Economic Security (CARES) Act, provided relief to the private sector, including loans to cover businesses' fixed costs, unemployment benefits, and temporary employment schemes. The program helped maintain demand for services provided by Palau's SOEs.

Figure 28: Net Profit 2015-2020 ($ million)

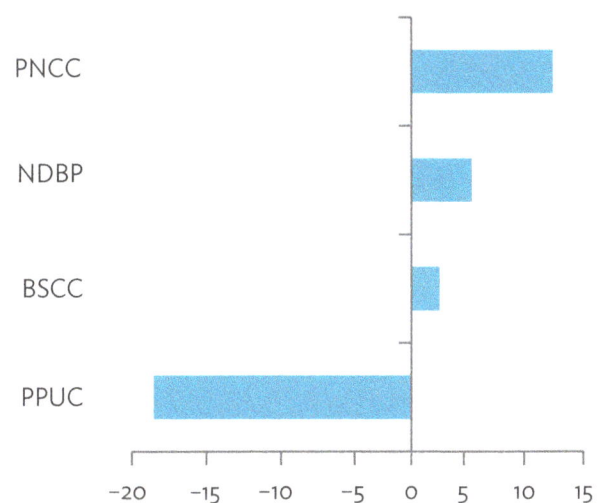

PNCC = Palau Public Utilities Corporation, NDBP – Palau National Development Bank, BSCC = Belau Submarine Cable Corporation, PPUC = Palau Public Utilities Corporation.
Source: State-Owned Enterprise audited financial statements

Three of the four SOEs were able to maintain the same level of core revenue in 2020 as in 2019. Only PNCC suffered a contraction in revenue, partially because of reduced visitor arrivals. The National Development Bank of Palau (NDBP) responded to COVID-19-affected customers by restructuring loans or repayment schedules, and increased its provisioning for doubtful loans by $2.5 million. NDBP stepped up its Housing Development Loan Program and received $5.7 million in related capital contributions from the government. ROA dropped from 4% in 2019 to –1% in 2020 because of increased doubtful loan provisioning, and NDBP expects 2021 to be equally challenging.

The government provided $2.1 million in operating subsidies to PPUC in 2020, up from $0.5 million in 2019. These subsidies were to cover the costs of delivering power, water, and wastewater services to low-income and unemployed residents, and were only partially a response to COVID-19. PPUC provided some customers additional relief by delaying disconnections and waiving fees and penalties for late payments. PNCC and Belau Submarine Cable Corporation (BSCC) neither sought nor received direct government support in 2020.

Palau's new government, which took office in January 2021, has accelerated SOE reform efforts to ensure that COVID-19-related support provides sustained benefits. These reforms, supported by a range of development partners (ADB, the World Bank, the United States, and Japan) involve all four SOEs and range in scope from infrastructure modernization to corporate governance, risk management, and policy and regulatory measures to place SOEs on a more commercial footing.

4. Legal, Governance, and Monitoring Framework

Palau's SOEs are statutory corporations regulated by their respective establishing legislation, which set objectives, governance arrangements, planning, reporting, and disclosure requirements. Only two of the SOEs (PPUC and BSCC) have an explicit objective in their establishing legislation to operate as commercial businesses and recover their costs of capital. Director qualification/disqualification varies for each SOE, but the appointment process is reserved for the President, who is given this power by the Constitution. While most SOEs are required to submit audited accounts to the auditor general, President, and Olbiil Era Kelulau (OEK), Palau National Congress, there is no consistent requirement to prepare corporate planning documents, nor to submit any information to the Ministry of Finance, apart from PPUC's monthly subsidies report.

There is no mechanism for monitoring the performance of the SOE portfolio apart from the presentation of each SOE's financial accounts to the OEK. An effective monitoring process would broadly track SOE businesses from their planning through to their results and performance assessment, with a view to protecting the government's shareholding interest in each SOE. This shareholding interest is focused on the SOE's medium- to long-term sustainability,

encouraging improved operational and financial performance and ensuring it covers its cost of capital.

Executive Order 465 of 2021 reaffirmed Palau's National Policy for SOE Governance, first issued in 2014. The policy established a common commercial objective for all SOEs, designated the minister of finance as the responsible minister for SOEs, and called for the establishment of an SOEMU to support the minister in monitoring SOE performance. This function would be supported through robust director selection, corporate planning, and reporting processes, and would complement the ministry's broader fiscal risk management responsibilities.

The ministry has established an SOEMU and prepared a SOE Bill to support the implementation of the policy. The bill will enable the ministry to undertake its SOE monitoring role, and create a framework for improved SOE financial performance. As is the case in other Pacific jurisdictions, noncommercial SOE activities would be identified, costed, and financed to improve service delivery and accountability. The bill's enactment should be prioritized in 2023.

5. SOE Reform Highlights 2015–2022

The issuance of the National SOE Policy in 2014 formalized the government's intention to place SOEs under a clear commercial framework. While the policy was preceded by the 2013 merger of the power and water utilities into PPUC (which is required to recover all its costs), these commercial objectives were also made explicit in the 2015 establishment of BSCC. In the absence of legislation to implement the policy, however, the Ministry of Finance has been unable to exercise its intended ownership monitoring powers.

PPUC undertook a series of corporate planning and governance reforms in 2021 and 2022, all of which were fully consistent with the SOE Policy. These reforms included the preparation of a 3-year corporate plan, a SCI, and a director selection process that it expects to implement under a future SOE law. Key to PPUC's commercial success will be its ability to charge tariffs enabling full cost recovery, and the establishment of a transparent mechanism to deliver tariff subsidies mandated by the government.

- The primary objective of each state-owned enterprise (SOE) is to operate as a successful business and, to that end, to recover all of its costs, including its costs of capital.

- Community service obligations delivered by SOEs shall be transparently identified, costed, contracted, and funded.

- The minister of finance will have financial and operational oversight over all SOEs.

- SOE directors will be selected based on their ability to bring the needed knowledge, skills, and experience needed by the SOE to meet its primary objective.

- Elected and appointed public officials shall not be appointed to SOE boards, and the number of civil servants will be limited.

- The boards of the SOEs shall prepare an annual statement of corporate intent. A monitoring unit will be established within the Ministry of Finance to monitor the performance of the SOEs and support the minister of finance.

Source: Pacific Private Sector Development Initiative.

The issuance of the National Public-Private Partnership (PPP) Policy and Transparency Guidelines in 2021 should facilitate SOE engagement with the private sector. The policy requires rigorous cost-benefit analysis of proposed infrastructure investments to assess potential procurement as PPPs, and supports transparent, competitive procurement.

6. PPUC and Climate Change

PPUC is a vertically integrated power and water utility, reaching 96% of households in Palau. In 2020, 98.5% of the energy generated by PPUC was from diesel, with the remaining 1.5% from solar. PPUC's renewable energy goals are set by the Palau Energy Policy and Road Map, which calls for 45% renewable energy generation by 2025. PPUC is working to achieve this goal with the establishment of solar IPPs, the first of which is a 20MW project awarded in 2020 and expected to be commissioned in 2023. This project will increase Palau's total installed generation capacity from 34MW to 54MW.

The main drivers of PPUC's renewable targets are to reduce energy generation costs, reduce emissions, achieve energy independence by reducing reliance on imported fuel, and comply with government-established renewable energy targets. The 2015 Energy Act encourages the use of renewable energy, but provides neither financial incentives nor penalties, so is not a major driver of PPUC's transition. Palau's regulatory framework allows net metering, with customers incentivized to install solar generation capabilities. However, the rollout has been hindered by the high cost of imported solar panels. In 2020, PPUC credited an estimated 2% of its revenue to net metering customers. PPUC's efforts to meet its own renewable energy goals are hampered by its weak financial position and inability to recover its costs of production.

Palau is considered highly vulnerable to extreme weather events, and three major typhoons in 2012, 2013, and 2021 caused an estimated $3.7 million of damage to PPUC's infrastructure. PPUC has made improvements to the electrical grid and substations, making them more resilient to the effects of climate change, but has not had sufficient financial resources to make similar investments in its water and wastewater assets. In 2021, PPUC developed a detailed corporate plan that disclosed the business risks posed by climate change.

PPUC's profitability has been volatile over the 2015–2020 period, and largely driven by tariff regulations and offsetting subsidy arrangements. PPUC received a total of $12 million in operating subsidies and grants from 2015 to 2020, and a further $19 million in capital contributions, yet generated a cumulative net loss of $18 million for the period. Caps on tariffs have constrained PPUC's ability to recover costs. Most recently, the Republic of Palau Public Law 10–42 restricted PPUC's ability to increase its electricity tariffs to recover increases in fuel costs during FY2020. COVID-19-induced supply chain disruptions delayed the parts, equipment, and expertise needed to maintain infrastructure and progress renewable energy projects in 2020–2021.

Figure 29: Palau Public Utilities Corporation Profitability 2015–2020

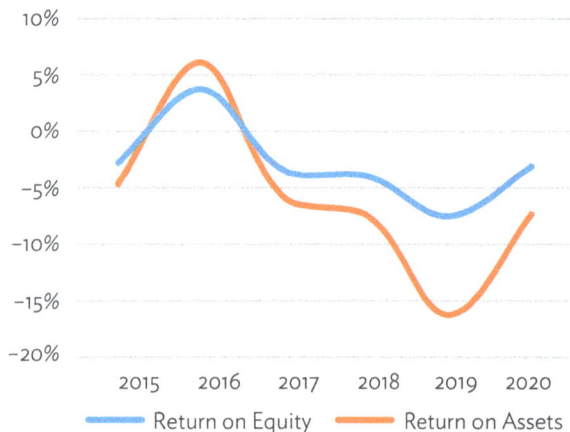

Source: Palau Public Utilities Corporation audited financial accounts

Reforms undertaken in 2021 have established an independent regulatory process for setting cost recovery electricity tariffs for PPUC. This should enhance transparency and reinforce PPUC's commercial mandate, placing it on a more sustainable footing. Other notable reforms include the strengthening of the Palau Energy Authority to serve as the regulator, and the introduction of robust corporate planning and financial management systems at PPUC. PPUC intends to integrate a disaster-risk management process into its business planning to prepare for future climate change events.

E. Papua New Guinea

1. SOE Portfolio Composition and Contribution to GDP

The Government of PNG has majority ownership of 16 active SOEs, 12 of which have their assets held in the General Business Trust (GBT) and are under the oversight of Kumul Consolidated Holdings (KCH), the trustee of the GBT. The four other SOEs include KCH, two resource sector companies,[69] and the National Airport Corporation. This benchmarking study focuses on 12 of the SOEs under KCH's oversight, all of which are companies registered under the Companies Act 1997. Eight of these SOEs hold dominant market positions in their respective sectors, and five are

regulated as monopolies.[70] The composition of the portfolio has remained relatively unchanged since 2010, with one SOE added in 2014 (PNG Dataco)[71] and another in 2019 (Kumul Agriculture).

Figure 30: State-Owned Enterprise Portfolio Assets 2020 (K10 billion)

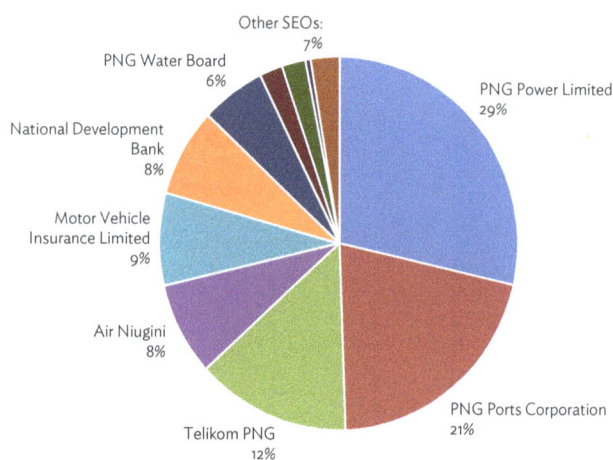

Source: State-Owned Enterprise financial accounts

The portfolio's overall contribution to GDP averaged 1.3% over the 2010–2020 period, substantially lower than the estimated 5%–8% of gross fixed investment in the economy controlled by the SOEs during this period. The contribution to GDP dropped sharply after the 2016 enactment of the Kumul Consolidated Holdings Act. The act transferred the oversight responsibilities for KCH and the SOEs to the National Executive Council (NEC), giving it responsibility for appointing SOE directors and approving corporate plans, and reducing KCH's role in administering the state's SOE ownership. This governance structure facilitated a deterioration in SOE performance (see Figures 31, 32), and a 50% increase in SOE-related debt, in some cases for projects not subjected to robust financial and economic viability assessments.

---

[69] Kumul Petroleum Holdings and Kumul Mineral Holdings.

[70] PNG Power Limited (PPL), PNG Water Board (WPNG), Eda Ranu, PNG Ports Corporation (PPCL), Air Niugini, Motor Vehicle Insurance Limited (MVIL), Telikom PNG, and Post PNG; the Independent Consumer and Competition Commission regulates PPL, WPNG, Post PNG, PPCL, and MVIL.

[71] PNG Dataco has not been included in the portfolio numbers presented here because of instability in the composition of its balance sheet from 2015 to 2020. During 2017–2019, the business was subsumed as part of the Kumul Telikom Holdings consolidation, and generated losses in 2015–2019.

Figure 31: State-Owned Enterprise Contribution to Gross Domestic Product (GDP) 2010–2020

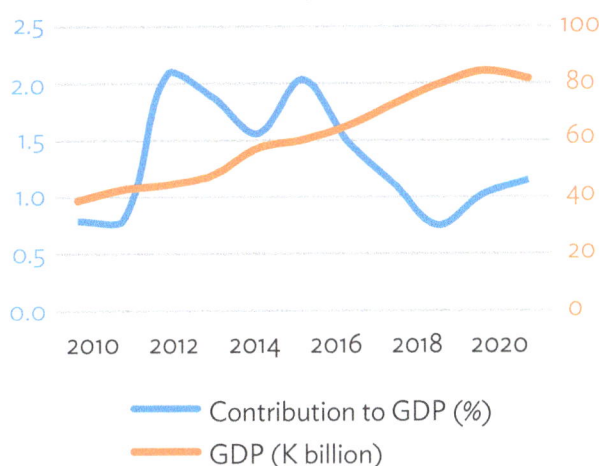

Contribution to GDP (%)
GDP (K billion)

Source(s): State-Owned Enterprise financial accounts, World Bank current GDP in Local Currency Unit data; PSDI analysis

Figure 32: State-Owned Enterprise Portfolio Profitability 2010–2020

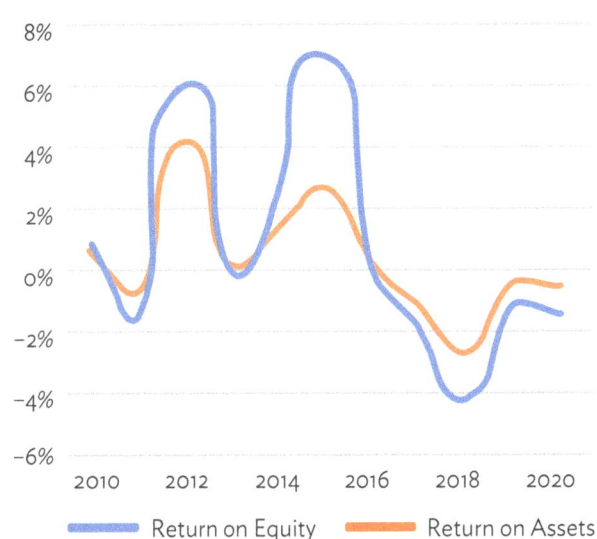

Return on Equity   Return on Assets

Source: State-Owned Enterprise financial accounts

2. SOE Performance 2010–2020

The portfolio's returns became negative in 2017 and are yet to recover. Surges in portfolio profitability in 2012 and 2015 were driven by asset revaluations in two SOEs: Motor Vehicles Insurance Limited (MVIL) quoted share investments in 2012 and PNG Ports assets in 2015. The latter added K300 million to PNG Ports' income for that year, an eightfold increase over 2014. Without this gain, 2015 net profit for the portfolio would have been K15 million and ROE 0% instead of 7%.

Three of the 11 SOEs had cumulative losses from 2010 to 2020 of K1.2 billion: Telikom PNG, Air Niugini, and Bmobile. Some 95% of the portfolio losses from 2010 to 2020 were incurred from 2016 to 2020. Average ROE and ROA were 2.5% and 1.2% respectively during the 2010–2015 period, and dropped to –1.5% and –0.8% in 2016–2020. This yielded an average ROE and ROA of 0.6% and 0.3% respectively for the period 2010–2020. From 2016 to 2020, 92% of the profits of the portfolio have come from two SOEs: PNG Ports and MVIL. In the case of MVIL, profitability was driven by the rising value of its quoted share investments, most notably Bank of South Pacific, in which it held a 6.67% stake in 2020.

As portfolio profitability deteriorated from 2015 to 2020, SOEs took on an unsustainable amount of new debt. An estimated K4.3 billion in new projects were approved, generating an additional K3.2 billion in debt, 44% of which was taken on directly by the SOEs at commercial rates.[72] At the end of December 2020, SOE (including KCH) interest-bearing debt totaled K5.15 billion, of which K1.85 billion was owed to commercial lenders and K3.65 billion to concessional lenders via Treasury. Some SOEs were already in technical default as of 2019, before any of the K3.65 billion in debt held by Treasury is transferred to them. Asset utilization in the SOE portfolio dropped from 43% in 2015 to 28% in 2020.[73]

The precarious financial position of the portfolio led the new minister of SOEs to launch a major SOE restructuring program in mid-2019. The objective of the program is to place the SOEs on a more sustainable financial footing by reducing debt, interest, and operating costs, and strengthening key governance arrangements. It came just as the PNG economy entered the COVID-19-induced economic downturn.

3. Preliminary 2021 Results

As of November 2022, five SOEs had audited financial accounts available for fiscal year 2021.[74] This subset of SOEs generated a consolidated ROE of 2.5% and consolidated ROA of 1.5% for 2021, compared to 5% ROE and 3% ROA in 2020. PNG Ports, Telikom PNG, and Water PNG saw revenue growth in 2021, while MVIL and NDB

---

72 This includes a K225 million loan to Kumul Consolidated Holdings to finance a dividend payment to the state in 2015.
73 Asset utilization measures the amount of revenue generated by each kina of assets in the portfolio; a ratio of 28% means that each kina of asset value in the portfolio produced 0.28 kina of revenue.
74 PNG Ports, PNG Water Board, Motor Vehicle Insurance Limited, National Development Bank, and Telikom PNG.

revenues declined. In the case of NDB, additional loan impairments resulted in a net loss of K36 million for the year. Water PNG generated negative returns for the first time since 2002 due in part to its amalgamation with Eda Ranu in April 2021.

4. COVID-19 Impact and Response

While most SOEs managed to maintain services and revenue levels in 2020, Air Niugini was hit particularly hard by the closure of international borders and reduced demand for domestic travel. Air Niugini's revenue dropped 43% while its operating expenses declined by 32%, resulting in a K97 million loss for the year. MVIL's profitability dropped 37% as the market value of its investments fell. Conversely, Water PNG and Eda Ranu improved their profitability in 2020 compared to 2019. Eda Ranu's operating expenses dropped 22% as it no longer paid concessionaire facility fees, while PNG Power maintained similar revenue levels to 2019 alongside a substantial reduction in fuel costs.

The government's ability to finance SOE restructuring measures was constrained by its own budgetary deficit and the requirements of COVID-19-related public sector spending. Support provided to Treasury by donor partners in 2020 was used in part to clear some government arrears with SOEs and to recapitalize Air Niugini, but most SOEs facing chronic operating losses have weathered 2020 without direct financial relief. Several SOEs have accumulated substantial tax arrears with the Internal Revenue Service, while accumulated receivables from the government stood at K365 million in the first half of 2021. The sale of some non-core assets is proceeding, but the largest and most valuable non-core assets (residential properties, land leases, and shares in Bank of South Pacific)[75] have yet to be monetized.

The prolonged effect of the economic downturn is likely to be felt for several more years by Air Niugini, and will complicate revenue growth for other SOEs facing chronic operating deficits. In the telecommunications sector, the merger of Telikom PNG and Bmobile will eliminate costly competition between their mobile businesses, but competition from the dominant player (Digicel/Telstra) and the new entrant (ATH) will continue to put pressure on prices and require ongoing investments. Debt relief remains urgent, while a broader strategy for the state's commercial role in the sector is yet to be determined. Similarly, PNG Power is seeking to reduce its costs of energy while addressing the vulnerabilities in its transmission and distribution network.

The 2020 SOE Ownership and Reform Policy signalled the government's intent to overhaul SOE governance arrangements, restoring KCH to its SOE oversight role and establishing a skills-based SOE director nomination process. The purpose of the policy is to create a more robust and de-politicized system of SOE governance, where boards are held accountable for SOE results and granted the independence to pursue commercial outcomes. While the minister for state enterprises remains responsible for safeguarding the government's investment in SOEs, the policy returned the responsibility for direct SOE oversight to KCH from NEC. This, together with enhanced transparency, should enable the SOEs to develop and implement restructuring strategies more quickly, free from direct political interference.

Box 8: Kumul Consolidated Holdings (KCH) Authorisation (Amendment) Act

Key features of the 2021 KCH Amendment Act:

Corporate planning:

- Three-year corporate plans to replace annual operating plans and include the identification and costing of community service obligations.
- Statements of corporate objectives (SCOs) to summarize key performance indicators.
- Corporate plans and SCOs must be reviewed by KCH and the minister before being submitted to the National Executive Council for approval.

State-owned enterprise (SOE) director nomination and evaluation:

- SOEs to develop and maintain board profiles.
- Terms of reference for board vacancies to be based on skills, experience, and diversity targets, and prepared jointly between SOE board and KCH.
- Women fulfilling terms of reference requirements systematically shortlisted.
- Minister presents KCH-recommended candidate to the National Executive Council for appointment
- Minister to review performance of KCH and SOE directors and boards every 2 years.

Disclosure:

- SOE SCOs to be publicly available.
- KCH and SOE audited financial statements to be prepared within 3 months of financial year end, and made available on the respective KCH/SOE websites after their presentation to Parliament.

Source(s): KCH Authorisation (Amendment) Act.

---

[75] The combined shareholding of 24.7% in Bank of South Pacific held by Kumul Consolidated Holdings and Motor Vehicle Insurance Limited had an estimated market value of A$470 million in December 2020.

5. Legal, Governance, and Monitoring Framework

KCH is a 100% state-owned statutory corporation mandated to hold all government-owned commercial (non-resource) assets in the GBT, and to manage them for economic, environmental, and social value. KCH is headed by a managing director and reports to NEC through the minister for state enterprises on matters relating to the KCH Act. The minister is also responsible for overarching policy matters related to the SOE portfolio.

Ministerial oversight of the SOE portfolio is complicated by the practice of conveying political responsibility for selected SOEs to sector ministers.[76] The designations under Section 148 of the Constitution convey political responsibility, rather than powers of control or direction, to ministers. However, the practice has led to confusion and dysfunction within SOE boards and senior management when receiving conflicting instructions from the minister of state enterprises, KCH, and the minister with political responsibility. This practice persisted through 2019 and 2020 when PNG Power, Telikom, MVIL, and NDB were placed under the political oversight of sector ministers.

Under the Kumul Consolidated Holdings Authorisation (Amendment) Act, passed by Parliament in August 2021, KCH has regained powers to monitor SOE performance, manage the SOE director nomination process, and review SOE corporate plans. While NEC retains power to make SOE director appointments and endorse SOE corporate plans, they do so upon the recommendation of KCH and the minister of state enterprises. In this way, the amendment is intended to reduce political influence on SOE decision-making that will strengthen accountability and commercial outcomes. Other important features of the amendments will strengthen the corporate planning process, increase public disclosure, and promote gender diversity on SOE boards.

Many SOEs provide CSOs for which they receive no compensation from the government. For the regulated SOEs,[77] these CSOs are financed through cross-subsidies recognized in regulatory contracts with the Independent Consumer and Competition Commission, but these provide few efficiency incentives or opportunities for third party service providers. The KCH Amendment Act requires SOEs to identify and cost their CSOs in their corporate plans and statements of corporate intent, as outlined in the government's 2012 CSO Policy. This is an important step towards securing direct financing for CSOs from the government budget, and competitively tendering some CSOs to private providers.

The government has made a commitment to progressing competitive neutrality for its SOEs under the GBT. NEC endorsed a National Competition Policy in 2020, calling on SOEs to achieve a commercial rate of return and face the same treatment as private firms in relation to taxes, regulation, and debt. Currently, a number of SOEs enjoy tax exemptions, and almost all source some of their debt financing on concessional terms from the government. The successful implementation of these competitive neutrality principles will remove the benefits received by SOEs by virtue of their state ownership. The costs of state ownership to the SOEs, conversely, will be addressed through the full implementation of the KCH Amendment Act and its protections from noncommercial and/or politicized decision-making.

Private sector involvement in service provision is inadequately enabled through PNG's current legislative and policy settings, so agreements are concluded on an ad hoc basis, with no consistent approach to assessing and negotiating commercial arrangements. In 2019, PNG was ranked equal 68th of 69 economies for its capacity to implement sustainable and efficient PPPs.[78] In 2014, the government passed the Public–Private Partnership Act 2014 (PPP Act) that provided for the systematic and transparent analysis of proposed infrastructure investments, and their competitive tendering. While the PPP Act was gazetted in early 2018, implementation stalled until the new government incorporated it into the 2019 SOE reform program.

6. SOE Reform Highlights 2015–2022

The first legislative reform regulating SOE performance in the 2015–2022 period was the amendment of the KCH Act in 2015,[79] which largely had a negative effect. The adoption of the SOE Ownership and Reform Policy in 2020, and subsequent amendments to the KCH Act in 2021, are intended to reverse these negative effects by establishing a

---

[76]   Under section 148 of the Constitution, the Prime Minister has the power to name the ministers with political responsibility over all departments, sections, branches, and functions of the government.

[77]   PNG Power Limited, PNG Ports Corporation, Post PNG, PNG Water Board, and Motor Vehicle Insurance Limited.

[78]   The Economist Intelligence Unit. 2019. *Measuring the Enabling Environment for Public–Private Partnerships in Infrastructure.* London. The study was commissioned by the Inter-American Development Bank, European Bank for Reconstruction and Development, and the Millennium Challenge Corporation.

[79]   Certified in 2016.

more robust governance framework. The new legislation's success will rest on continued political commitment to its principles, liberating the SOEs to focus on achieving commercial outcomes.

A PPP Amendment Act was endorsed by Parliament in 2022. The Amendment Act and its regulations will facilitate the establishment of a small PPP center and the implementation of a rigorous PPP project development process. This will add capacity to SOEs that are seeking partnerships with the private sector, improve VFM outcomes, and reduce risk for private investors through predictable and transparent tendering.

Three PPP transactions have been implemented since 2015, only one of which was subject to a competitive tendering process. The first was the container terminal concession for the ports of Lae and Port Moresby, signed by PNG Ports and International Container Terminal Services Inc. in 2017, which has more than doubled cargo handling productivity in its first 3 years. The second and third PPP transactions were IPP contracts signed with PNG Power in 2019—one with Niupower and the other with Dirio Gas and Power. Neither of these were subject to a competitive tender process. In the absence of the PPP Act, the PPP project development process and treatment of unsolicited proposals remains largely unregulated, raising the risk of uneconomic outcomes for SOEs.

The long-awaited restructuring of PNG's power and telecommunications sectors was largely hampered by the SOE governance arrangements in place before the 2021 KCH Amendment Act. The government now has an opportunity to finalize these reforms, enhancing competition in the telecommunications sector through regulation and partial privatization, and in the power sector through the establishment of the new regulator (National Energy Authority) and the introduction of more private participation in service provision.

7. PNG Power Ltd and Climate Change

PNG Power Ltd (PPL) is a vertically integrated state-owned power utility, which produces about 50% of the power consumed in PNG. The balance is produced by the private sector, including for their own captive

Figure 33: Fuel Mix for PNG Power Limited Energy Produced

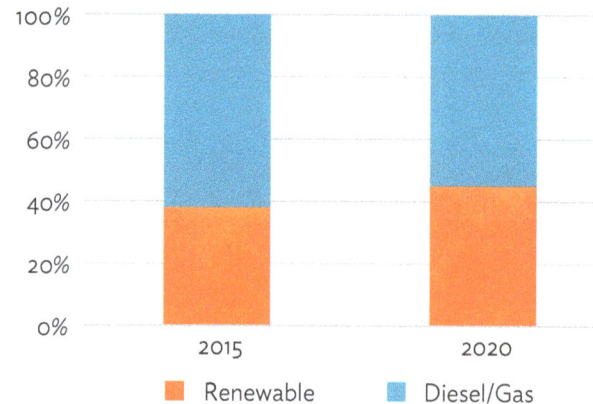

Source: International Renewable Energy Agency, PNG Energy Profile 2020.

use. Currently, PNG has one of the lowest per capita consumptions of electricity in the world, with an estimated 20% of the population having access to grid or off-grid electricity. PPL is licensed under the Electricity Industry Act to generate, transmit, distribute, and sell electricity throughout PNG, with the exclusive right to supply small customers (<10MW load) within 10 kilometers of its network. It is subject to the government's climate change mitigation and adaptation policies, most notably the goal of facilitating access to electricity for 70% of the population, and becoming fully carbon neutral by 2030.[80] Some 45% of PPL's power supplied to the grid in 2020 was from renewable sources, the second highest in the Pacific (behind EFL at 64%), and is likely to increase in the coming decade as additional hydropower and solar generation capacity is added to the grid to displace diesel. It is estimated that, of the 1,037MW of installed generation capacity in PNG in 2019, 32% was from renewable sources, of which 77% was hydro, 17% geothermal, 5% biomass, and 1% solar.[81]

PPL's renewable energy targets are primarily driven by cost considerations, and guided by its Least Cost Power Development Plan. This plan calls for the rehabilitation of existing hydro and the replacement of diesel with domestically produced natural gas in the near term, and the addition of new hydro and other renewables as demand increases in the medium term.[82]

---

[80]   Electricity Industry Policy of 2011, Development Strategic Plans (2010–2030), Vision 2050, and Second Nationally Determined Contributions. 2020. https://www4.unfccc.int/sites/NDCStaging/pages/Party.aspx?party=PNG.

[81]   Asian Development Bank. 2021. *Pacific Energy Update 2021*. Manila.

[82]   World Bank. 2020. *PNG Power Least Cost Power Development Plan Update 2020*. Port Moresby, PNG.

Challenges to achieving these renewable energy targets include PPL's weak financial position, with effective tariffs below the cost of production.[83] High input costs, equipment failures, and low collection rates, in particular with government clients, also contribute to chronic losses. PPL estimates it incurs close to K1 billion of CSO costs per year, for which it receives no compensation from the government. These include the requirement to operate and maintain nonprofitable power grids, government directives to enter into noncommercial power purchase agreements, and tariffs which have not increased since 2013 despite rising costs. PPL estimates the excess payments attributable to two of its noncommercial PPAs, executed without competitive tender at the direction of the government, at K120 million per year.

PPL has experienced substantial fluctuations in profitability since 2017, when sharp increases in input and operating costs outpaced revenue growth. Fuel and IPP costs went from 36% of revenue in 2016 to 44% in 2017, and 50% in 2018. Plans to restructure PPL, secure CSO payments, and renegotiate existing PPA agreements are currently on hold, but a recovery strategy will be needed to place the utility on a sustainable financial footing and facilitate planned investments to increase renewable energy use. PPL also intends to subject future IPP contracts to robust risk assessments and competitive tendering, as will be required under the PPP Act. The first such competitive IPP tender, for a 5MW solar plant on the Gazelle grid, was launched in 2021 and is expected to reduce PPL's reliance on diesel generation. In 2022, high fuel costs drove the costs of generation on the Gazelle grid to more than three times the projected offtake price under the solar IPP.

PPL recognizes the physical risks to its assets posed by climate change, but has not yet invested in mitigation measures. While future investments in transmission and generation infrastructure are expected to incorporate climate-resilient designs, existing infrastructure remains vulnerable. PNG is prone to natural hazards such as earthquakes, volcanic eruptions, tsunamis, landslides, and storm surges. PNG is a member of the Vulnerable Twenty (V20) group,[84] and is ranked 149th out of 181 countries by an index reflecting its vulnerability to, and readiness for, climate change-related challenges.[85] PNG supports the

Sendai Framework for Disaster Risk Reduction and the Paris Climate Agreement. PNG's annual carbon emissions are low. Excluding land-use change and forestry emissions, these were estimated at 7.8 metric tons of carbon dioxide in 2018, which ranks PNG 116th from 222 countries.[86]

Figure 34: PNG Power Limited Profitability 2010–2020

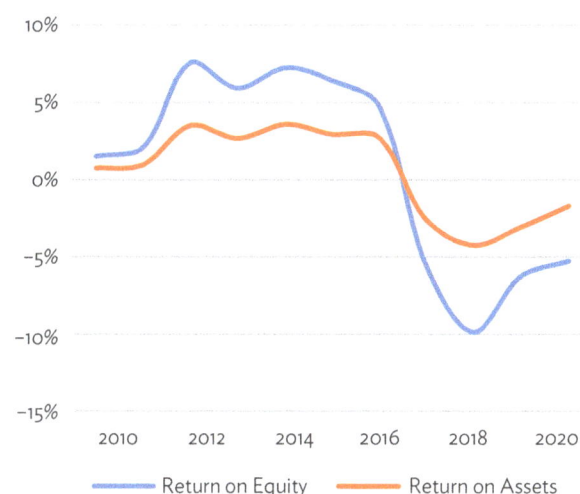

Source: PNG Power Limited annual financial accounts

F. Samoa

1. SOE Portfolio Composition

Samoa's SOE portfolio comprises 14 entities engaged in a diverse range of activities including transport, utilities, subsidized housing, postal services, banking, land development, and trustee services. The six largest SOEs represented 84% of total portfolio assets in 2020, but only 26% of total revenue. The portfolio is large, with total assets of ST1.77 billion, but contributed only 3% per year to GDP in the decade 2010–2020.

---

[83] Financial data received a qualified audit report in 2017 and 2018. The 2015, 2016, 2019, and 2020 numbers are unaudited and based on PNG Power Limited management accounts.

[84] The V20 group of finance ministers, established in 2015 as part of the Climate Vulnerable Forum, is a dedicated cooperation initiative of economies systemically vulnerable to climate change. It works through dialogue and action to tackle global climate change.

[85] Notre Dame Global Adaptation Initiative. https://gain.nd.edu/our-work/country-index/rankings/.

[86] Global Carbon Atlas. http://www.globalcarbonatlas.org/en/CO2-emissions.

2. SOE Performance 2010–2020

The composition of the SOE portfolio has remained reasonably stable since 2006, making analysis of performance over time meaningful.[87] Three SOEs have been privatized since 2006: Samoa Broadcasting Corporation in 2008, SamoaTel in 2009, and Agricultural Stores Corporation in 2015. The portfolio's returns slightly improved in the period 2010–2019, but both ROE and ROA dropped to 0% in 2020. Annual returns have been volatile over the decade to 2020, but the average portfolio ROE and

Figure 35: State-Owned Enterprise Portfolio Assets 2020 (ST1.8 billion)

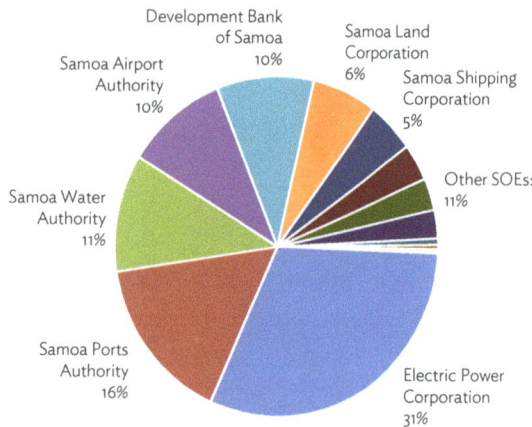

Source: State-Owned Enterprise audited financial accounts

ROA were 0%, increasing to 1% and 0.5% respectively in the period 2016–2019. Portfolio returns were adversely impacted by Polynesian Airlines (PAL), which recorded cumulative losses of ST57.8 million for the 2018–2020 period, the latter 2 years impacted by a measles epidemic and COVID-19 travel restrictions.[88] If PAL's results are excluded, average ROE and ROA would have been 3% and 1% respectively over the period 2017–2020 (Box 9).

While the improvement over the past decade has been modest, and the volatility in 2017–2018 is because of movements in a small number of large SOEs, the strengthened governance and improved ownership monitoring practices since 2015 appear to be effective. These follow the establishment of the Ministry of Public Enterprises (MPE) and amendments to the SOE Act in 2015.

Figure 36: State-Owned Enterprise Portfolio Profitability 2010–2020

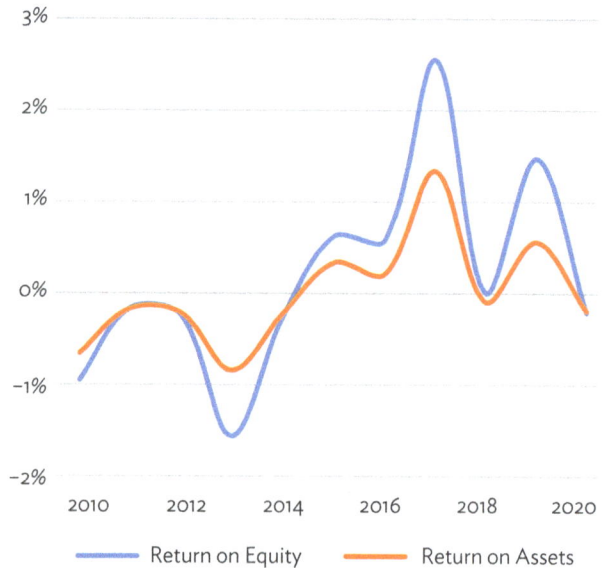

Source: State-Owned Enterprise audited financial accounts

Prior to this, ownership monitoring was undertaken by a unit within the Ministry of Finance.

Historically, Samoa's SOEs have carried the burden of informal and unfunded CSOs, adversely impacting profitability. Only two SOEs, Samoa Water Authority and EPC, are formally recognized as providing CSOs for which they are funded. But many other SOEs are directed to undertake noncommercial activities for which they are not compensated, such as the Development Bank of Samoa (DBS) providing subsidized rental for tenants in its head office building, and concessional loans; Samoa Post Limited maintaining district post offices where costs exceed revenue; and Samoa Housing Corporation providing subsidized rentals. The continued influence of senior ministers on SOE decision-making has increased the risk of unfunded CSOs to achieve social policy objectives, rather than the commercial objectives mandated under the SOE Act. Recognizing the significant adverse impact of informal CSOs on SOE profitability and operational efficiency, the MPE has drafted amendments to the SOE Act to clarify the definition of CSOs and strengthen the approval process.

[87] Two State-Owned Enterprises (SOEs) were reclassified as nonprofit in 2020, Land Transport Authority and the Casino Gambling Authority, and one was reclassified as mutual, Accident Compensation Corporation. An entity previously classified as nonprofit, Samoa International Finance Authority, was reclassified as an SOE in 2021, and one was created in 2011, the Unit Trust of Samoa (Management) Limited.
[88] Polynesian Airlines Limited's financial results for FY2020 are from unaudited accounts.

3. Preliminary 2021 Results

Samoa's SOEs[89] showed a decline in overall performance for fiscal year 2021 compared to 2020, consistent with the country's economic contraction. Portfolio revenue was down 12%, driving down consolidated ROE to 0.5% (from 2.6% in 2020) and ROA to 0.2% (from 1.3% in 2020). The Samoa Airport Authority (SAA) and DBS generated the largest losses due to the COVID-19 border closures.

The financial year ending 30 June 2021 proved to be a historically challenging year for aviation globally. SAA managed to reduce its overall costs by 10%, although it was unable to fully offset the 67% contraction in revenue. DBS was also adversely affected by the border closures, as 70% of its loan portfolio was exposed to the tourism sector. In response to COVID-19, DBS provided relief assistance to tourism providers by reducing loan and interest repayments. This in turn affected DBS' interest income (its main revenue source), resulting in an overall revenue contraction of about 41% compared to 2020.

4. COVID-19 Impact and Response

COVID-19, following a measles epidemic in 2018, has adversely impacted the financial and operational performance of many SOEs, but most notably PAL. The airline incurred losses of ST24.2 million in 2019 and ST17.8 million in 2020, and the government was required to provide a capital injection of ST35 million. Staff have been affected also, and salaries have been reduced by 50%. Other SOEs impacted include DBS and SAA; Samoa Shipping Services, whose major income source is commission revenue from Samoan seafarers serving on cruise ships; and Samoa Land Corporation, because of a 5-month rent-free period for tenants in its markets. The government has also used its SOEs to either provide or deliver a range of COVID-19 response measures (Box 10), in most cases without recognizing these additional costs as CSOs.

5. Legal, Governance, and Monitoring Framework

Samoa's SOE Act[90] establishes an effective SOE governance and monitoring framework. However, implementation has been erratic, mainly because of political meddling in SOE operations. This was exacerbated by the sector minister being one of the two shareholding ministers appointed by Cabinet. Prior to 2015, SOE monitoring was undertaken by a unit within the Ministry of Finance, which also limited its effectiveness. These issues were addressed in a 2015 amendment to the SOE Act creating an SOE minister responsible for all SOEs, and establishing the MPE in July 2015 to support the new minister.

All SOEs are required to produce 4-year corporate plans and publish a summary of the plan, called the statement of corporate objectives. SOEs report progress against plan targets quarterly to the MPE; the MPE then publishes summaries of the quarterly reports on its website. The MPE also produces a corporate plan covering its operations and

---

Box 9: Airlines are a Risky Business

Polynesian Airlines operates international and domestic routes. The international arm is marketed as Samoa Airways and commenced operations in November 2017; in the 3 years since, it has accumulated losses of ST51.5million. Cumulative losses have wiped out shareholders' funds which, by 30 June 2020, were ST17.9 million.

This is not the first time the government has owned a state-owned enterprise that is operating an international air service. Polynesian Airlines flew internationally until 2005 and was also loss-making—cumulative losses from 2002 to 2004 totaled ST48.6 million. In its final year of operation, the airline's annual losses equated to 20% of the government's budget.

In 2005, the government signed a joint venture with Virgin Australia to establish Polynesian Blue, which was 49% owned by the government, 49% by Virgin Australia, and 2% by private interests. While Polynesian Blue's accounts are not publicly available, it paid the government a ST5.7 million dividend in 2016. A post-transaction assessment was undertaken by the International Finance Corporation, which noted a 130% increase in inbound seat capacity, indirect tax collection from additional tourist arrivals of $1.86 million, and a positive fiscal impact of $6.9 million in the period 2005–2009. From 2005 to 2017, Polynesian Airlines only flew domestic and American Samoa services but was profitable, earning an average 6% return on assets over that period.

The government's experience with Polynesian Airlines shows the risks associated with airline ownership, in particular when facing international competition. The decision in 2018 to acquire a B737 MAX, the measles epidemic, and coronavirus disease (COVID-19) in 2020 created significant adverse consequences for Samoa Airways. The government would have been partially shielded from these consequences if it stuck with the joint venture. In June 2022, Samoa Airways formally ceased servicing international routes (except to American Samoa through Polynesian Airlines).

Sources: Polynesian Airlines audited financial accounts, Pacific Private Sector Development Initiative.

---

[89] As of November 2022, audited annual reports for fiscal year 2021 were available for 12 of the 14 state-owned enterprises surveyed in 2020. The two entities that were unavailable were Polynesian Airlines and Samoa Trust Estates Corporation.

[90] Public Bodies (Performance and Accountability) Act 2001 and Associated Regulations.

an annual report, reporting progress against plan targets. Both reports are published on its website, along with relevant policy and procedure manuals. Annual reports must be submitted to the MPE within 4 months of the end of the financial year. While SOEs have not achieved 100% compliance, there has been a steady improvement since 2015. Ten SOEs publish their annual reports on their websites.

Samoa's SOE director selection and appointment process is one of the most comprehensive within benchmarked countries. The process is codified in a schedule to the SOE Act and establishes an effective skills-based selection and appointment process. An Independent Selection Committee, comprising three members from the private sector, is primarily responsible for candidate selection, supported by the MPE. Three short-listed candidates are recommended to the responsible minister, who recommends a preferred candidate to Cabinet for appointment.

6. SOE Reform Highlights 2015–2022

Strengthening the legal, governance, and monitoring framework for SOEs has been the primary reform focus since 2015. In 2015, Cabinet approved the privatization of four SOEs: Samoa Post, Samoa Housing Corporation, Public Trust Office, and Agricultural Stores Corporation (ASC). The only transaction completed was the sale of ASC in 2016. In 2019, Samoa Ports Authority entered into 5-year performance based contracts with its existing stevedores. EPC has also pursued IPP contracts for solar power generation, adding 10MW of capacity, and completed a tender for 69 additional MW in 2022.

In 2022, the Ministry of Public Enterprises launched a review of its PPP Framework, seeking to integrate it more fully into the government's infrastructure project appraisal process. This reform should ensure that all PPP projects undergo the same rigorous project development process, including fiscal risk and value for money assessments.

The 2015 amendment to the SOE Act, the appointment of a responsible minister, and the establishment of the MPE drove a much-needed focus to address issues that were frustrating SOE reform. Initiatives included:

Box 10: COVID-19 Response Measures Delivered through State-Owned Enterprises

Development Bank of Samoa (DBS)
- Three-month rental holiday for tenants in DBS building.
- Two-month interest relief for all loans, and extension of 18 months for nonperforming loans.
- Government provided ST2 million package to support the bank's coronavirus disease (COVID-19) support.

Electric Power Corporation
- A 50% reduction in hotels' daily fixed rate for 3 months.
- A ST0.10 reduction in price of electricity for 6 months (Ministry of Finance funds ST0.07, and Electric Power Corporation funds ST0.03).

Samoa Airport Authority
- Three-month rent holiday for all businesses operating within the Faleolo Airport Samoa Housing Corporation.
- Three-month moratorium on clients' loan repayments, and interest rates reduced by 50% for 6 months.
- All residential tenants receive 1 month of free rental, and 50% reduction for 3 months.
- Commercial tenants receive 3 months' rent free.

Samoa Land Corporation
- Two-month rent holiday at all Samoa Land Corporation markets.
- A 15% reduction on lease rates for all tenants, 50% reduction on interest charged for all land sales, and 50% reduction in property lease rates.

Samoa Ports Authority
- Three-month rent holiday for all businesses operating on all wharves.
- Three-month refund on all stevedoring licenses.
- A 20% reduction on all wharfage fees for 3 months.

Samoa Water Authority
- A ST0.20 reduction in water rates for 6 months (compensated by Ministry of Finance).

Source: Ministry of Public Enterprises.

- **SOE categorization.** The SOE Act not only deals with SOEs, but also nonprofit and mutual corporate entities, labeled public beneficial bodies and public mutual bodies, respectively. In 2019, the MPE reviewed the legal and operational structure of each body to ensure that they were properly categorized, resulting in four reclassifications.

- **CSOs.** Simplified the definition of CSO, developed costing guidelines, and strengthened the application of the CSO framework.
- **Corporate plans.** In 2018, MPE undertook an audit of corporate plans and compliance with the legal framework, provided training on the development of corporate plans, and aided specific SOEs to improve their plans.
- **SOE director selection.** The director selection process and the functioning of the Independent Selection Committee have steadily improved since 2015. Amendments to the SOE Act have been drafted to codify the enhanced process, and clarify accountabilities.
- **Dividend and ROE targets.** In 2017, new minimum dividend and ROE targets were adopted. The minimum annual dividend was set at 35% of NPAT, and the minimum ROE at 7%.

Having strengthened the legal, governance, and monitoring framework, the government should again look at the portfolio and consider privatization of nonstrategic SOEs that are competing with the private sector, or where competition is possible. The three SOEs approved for privatization in 2015 could be early candidates.

7. Electric Power Corporation and Climate Change

Electric Power Corporation (EPC) is a vertically integrated power utility that produces and distributes about 87% of power consumed in Samoa. The power grid reaches 95% of the population. In fiscal year 2021, 45% of total energy produced was from renewable sources, up from 26% in 2015.

EPC is supporting the government's ambitious target of 100% renewable energy generation by 2025, and intends to use IPPs to achieve it. In fiscal year 2021, IPPs produced 11% of total generation and 86% of solar generation.

In 2019, EPC partnered with GridMarket to launch a tender for renewable energy production on an IPP basis. The resulting contract, for 69MW of solar and 196 MWh of battery storage, is awaiting government endorsement. The tender process—conducted outside of the PPP framework established by the PPP Unit—has highlighted the importance of undertaking fiscal risk assessments as an integral part of PPP project development.

Prior to 2015, Samoa's Cabinet could, and often did, appoint the sector minister as the responsible minister for SOEs operating within their economic sector. This created a conflict of interest, as the sector minister was required to make decisions on sector policy, regulation, and purchase of community service obligations, and also on matters impacting the commercial mandate and exercise of ownership rights of SOEs. This conflict was addressed in the 2015 amendment to the SOE Act which established a separate SOE minister as the responsible minister for all SOEs. However, some sector ministers were unwilling to relinquish the influence they held over the SOEs, and attempted to influence decision-making by holding regular policy meetings with SOE chief executive officers and submitting candidates for appointment to SOE boards direct to Cabinet. This confusion was compounded by conflicts between some SOE establishing acts and the SOE Act, where sector ministers retain some powers, such as tabling annual reports in Parliament. The Ministry of Public Enterprises is working with the government to eliminate these conflicts.

Source: Pacific Private Sector Development Initiative.

The main drivers of EPC's renewable targets are to reduce energy generation costs, reduce emissions, achieve energy independence by reducing reliance on imported fuel, and comply with the government's renewable energy commitments. EPC has prioritized its investment in climate-related resilience, based on past experience, and identified vulnerabilities. EPC is investing in ABC cable, which is thick and well insulated, for its low-voltage lines to counter against lightning strikes such as the one that caused a major power outage at its Fiaga plant in 2018. The cables are less likely to cause power outages if they meet tree branches and similar objects during strong winds, and will not cause electrocution if the cable drops to the ground (unless it is deliberately cut). EPC has also invested in laying cables underground as cyclone mitigation.

While EPC's profitability has been volatile over the past decade, it has generally been profitable. Changes in diesel prices have had a significant impact on costs, but this will reduce as renewables become a larger proportion of its generation mix. Unfunded CSOs also have negatively impacted financial performance. In addition to the formally recognized CSOs of street light installation and operation, for which EPC reports an annual shortfall of ST1 million, EPC has borne the cost of COVID-19 tariff reductions, estimated at ST9.6 million in 2021.

COVID-19 has had a significant impact on EPC, both operationally and financially. Reduced demand, driven mostly by the collapse of the tourism sector, reduced core operating profit after tax from ST11.5 million in 2019 to ST5.1 million in 2020 and ST2.8 million in 2021. Accounts receivables with large customers also increased, particularly those engaged in tourism-related activities, which requested extended time frames to pay bills. There were delays in the receipt of input materials because of shipping and transport disruption, which impacted EPC's ability to connect new customers and undertake upgrade and repair work on existing infrastructure. Lockdowns also limited the availability of technical staff to support operations.

Figure 37: Electric Power Corporation Profitability 2010–2021

Source: Electric Power Corporation audited accounts

G. Solomon Islands

1. SOE Portfolio Composition and Contribution to GDP

The Government of Solomon Islands has majority ownership of 11 active SOEs, six of which are included in this benchmarking analysis.[91] The portfolio is dominated by the power utility (SIEA) and the ports (Solomon Islands Ports Authority [SIPA]), which together represented 80% of total assets in 2020, and have contributed an average of 94% of the portfolio's net profits since 2015.[92] The composition of the portfolio has remained relatively unchanged since 2010, with one SOE added in 2017 (Solomon Islands Airport Corporation), although it is yet to be placed under the supervision of the Ministry of Finance. Two of the seven SOEs are limited liability companies registered under the Companies Act 2009 (Solomon Airlines and Solomon Islands Airport Corporation), while the others are statutory authorities. All SOEs hold dominant market positions in their respective sectors.

Figure 38: State-Owned Enterprise Portfolio Assets 2020 (S$3.1 billion)

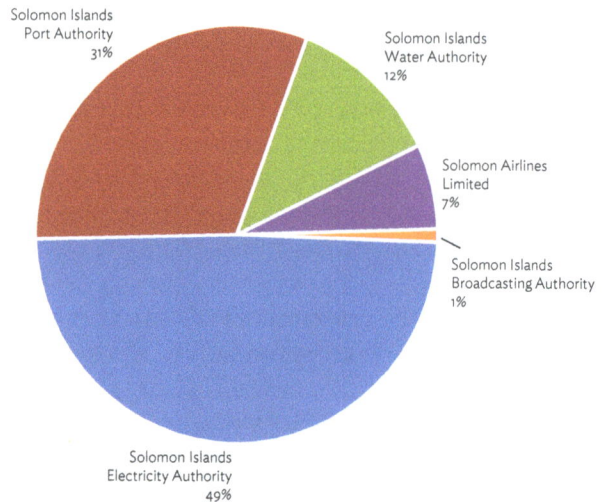

Source: State-Owned Enterprise audited accounts

91   Two state-owned enterprises (SOEs) are listed under the SOE Act 2007 but not included in this study: Commodities Export Marketing Authority (CEMA) and the Investment Corporation Solomon Islands (ICSI). CEMA is principally a regulatory agency without commercial activities, and ICSI is an SOE holding company. Two other SOEs, Solomon Islands Development Bank (recapitalized in 2018) and Solomon Islands Submarine Cable Company (established in 2016), fall under neither the supervision of the Ministry of Finance nor the regulation of the SOE Act 2007, and thus are excluded from this study. The Solomon Islands Airport Corporation is still operating under the Ministry of Communications and has not yet produced financial accounts.
92   The 2019 and 2020 figures do not include Solomon Islands Postal Corporation, accounts for which were not available as of November 2022.

The portfolio's overall contribution to GDP averaged 3% over the 2010–2020 period, lower than the estimated 16%–28% of gross fixed investment in the economy that it controlled in 2020. The portfolio's contribution to GDP, and overall profitability, improved dramatically after 2010 compared to the previous 8 years, when it was chronically loss-making.[93] The turnaround can be attributed to the financial restructuring of three of the largest SOEs, improved collections and tariff setting, and the implementation of the 2007 SOE Act and its 2010 regulations, which require SOEs to operate as commercial businesses.

Figure 39: State-Owned Enterprise Contribution to Gross Domestic Product (GDP) 2010–2020

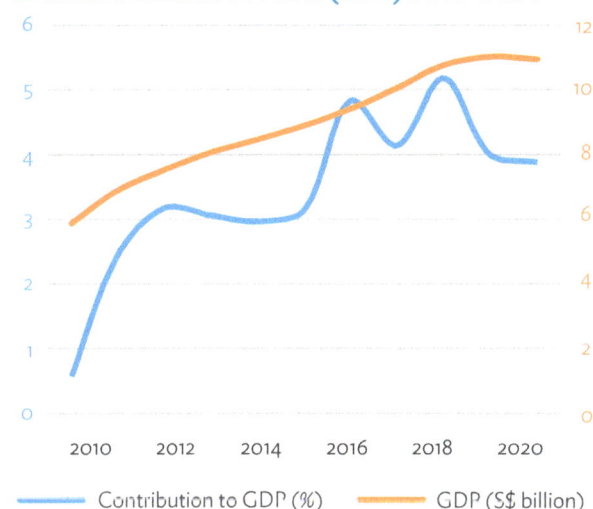

Sources: State-Owned Enterprise audited accounts, PSDI analysis

2. SOE Performance 2010–2020

The portfolio is the most profitable in this benchmarking sample, averaging a 10% ROE and 7% ROA over the 2010–2020 period. SIPA and SIEA generated more than 90% of the total profit of the portfolio over the 2010–2020 period, with Solomon Airlines (SAL) generating a net loss. Cost-based tariff adjustments and efficient collections have enabled SIEA to average a 9% ROE and 8% ROA from 2015 to 2020. The Solomon Islands Water Authority (SIWA), which has been unable to fully recover costs through its tariff, has received CSO payments to partially offset these costs, and a reduced average ROE of 5% and ROA of 3% to cover the balance. A surge in portfolio profitability in 2016 was largely due to sharp tariff increases at SIPA, leading to

a 75% increase in revenue. Barring SAL and the Solomon Islands Postal Corporation, most SOEs have benefited from asset grants which have kept debt levels to less than 30% of total assets, enabling them to absorb the high cost of this debt (7% average interest rate in 2010–2019). However, asset utilization[94] has been decreasing from an average of 87% between 2010 and 2014 to 43% between 2015 and 2020, signaling reduced productivity.

The government's implementation of CSO regulations yielded total payments of S$69 million from 2011 to 2020, or about 5% of total portfolio profits. While the budget set aside by the government for CSO payments each year is not defined by the CSO cost estimates, it is nevertheless an important contribution to offsetting CSO costs and increasing transparency. Because of government budget limitations, approved CSO funding is consistently lower than CSO cost estimates. This requires a majority of noncommercial activities to be funded by SOEs through an agreed lower rate of return. However, irregularity in the implementation of regulations has eroded their effectiveness, as some SOEs have received differing payments from year to year, despite incurring similar CSO costs. Increased predictability in the implementation of the regulations, and the robust contracting framework which they support, would improve the commercial results of the SOEs and the government's ability to deliver CSOs in a sustainable and cost-effective manner.

3. COVID-19 Impact and Response

While most SOEs managed to maintain services and revenue levels in 2020, SAL and the Solomon Islands Airport Corporation were particularly damaged by the closure of international borders and reduced demand for domestic travel. SAL's operating revenue dropped 46% while reducing operating expenses 37%, resulting in a S$31 million loss for 2020, despite a CSO payment of S$4 million and government grant of S$20 million. SIEA suffered a 7% drop in core revenue and 19% drop in profit, while SIWA was able to grow core revenue by 3% despite reduced demand from commercial customers. SIPA managed a 6% increase in core revenue in 2020 compared to 2019, but a 27% increase in administrative expenses reduced its overall profit margin by S$33 million compared to 2019. Most SOEs have delayed the implementation of existing capital projects as border closures restrict the flow of goods and expertise, and have postponed future projects in an effort to preserve cash.

[93] The portfolio's average contribution to gross domestic product and return on assets from 2002 to 2009 was 0% and -4% respectively.
[94] Measures the amount of revenue a firm is able to generate from its assets.

The government's response to COVID-19 included a request to SIEA, SIPA, and SIWA to reduce tariffs, with SIWA receiving a S$5 million payment to offset this cost. SIEA and SIPA were asked to finance the reduced tariffs through a lower return on investment. SAL received a S$20 million economic stimulus package grant and a S$4 million CSO payment. SIWA received additional donor grants and contributions totaling an estimated S$11 million in 2020, enabling it to generate a net profit of S$17 million for the year.

The prolonged effect of the economic downturn is likely to be felt for several more years by SAL, which was in a weak financial position before COVID-19. After generating accumulated losses of S$70 million from 2012 to 2016, which wiped out SAL's shareholder equity, new cost-cutting measures allowed it to briefly return to profitability in 2017–2018 before generating losses again in 2019–2020. SAL remains in a precarious financial position, with a negative net equity (even after the government's S$20 million grant in 2020). High levels of current liabilities (59% of total debt) are not covered by current assets, with a quick ratio averaging 0.18 from 2017 to 2020. In addition to the recovery plans developed by SAL's management and board, further financial restructuring will be required to enable SAL to withstand future shocks.

4. Legal, Governance, and Monitoring Framework

The SOE Act and its supporting regulations establish a robust framework for commercially managing SOEs.[95] It requires the SOEs to operate profitably, imposes a rigorous director selection and appointment process, defines corporate planning and reporting requirements, and establishes a process for the transparent identification, costing, and financing of CSOs. Adherence to the act has varied under different governments over the 2010–2020 period, particularly regarding the director selection and appointment process and the implementation of CSO provisions.

The shared responsibility for SOE oversight by the accountable ministers, one of which is the line minister and the other the minister of finance, has complicated governance and weakened some of the act's core features. This dual oversight model blurs the distinction between the regulatory and ownership responsibilities of the two ministers, and introduces conflicts for the line minister, who

must reconcile sector policy with the SOE's commercial mandate. It is also inconsistent with global trends, which are moving increasingly towards centralized ownership monitoring structures.

The government intends to strengthen and consolidate SOE oversight arrangements through forthcoming amendments to the SOE Act and regulations. These amendments will also formalize the role of the SOE Unit of the Ministry of Finance in the director selection process, establish a director and board performance evaluation process, and support increased gender diversity on SOE boards.

Many SOEs continue to struggle with corporate planning and reporting. With the exception of the largest SOEs, SOEs struggle to prepare meaningful statements of corporate objectives (SCOs) and corporate plans, and to submit their audited annual accounts within 3 months of a financial year.[96] In 2019 and 2020, none of the SOEs submitted their audited annual accounts within the required time, in part because of COVID-19-related delays and the late appointment of the new auditor general. As of November 2022, SIEA was the only SOE which had completed its 2021 audited accounts. SCOs often do not include the required content. The SOE Unit within the Ministry of Finance provides oversight and reports on SOE performance to the minister of finance, but lacks authority to enforce adherence to the SOE Act.

Despite its limited resources, the SOE Unit has managed to review corporate plans and SCOs, review CSO cost estimates, provide briefings for the minister, and maintain an online portal to publish SOE financial results. Solomon Islands is one of only three countries in this benchmarking sample (with Tonga and Samoa) whose SOEMU is reporting on portfolio performance. This is an important accountability practice that requires ongoing resources (and SOE compliance with statutory reporting obligations) to ensure that the information is current and user-friendly.

5. SOE Reform Highlights 2015–2022

The government adopted an SOE Ownership Policy in 2018 to guide future investment in SOEs, including through partnerships with the private sector. It committed the government to review the rationale and VFM of its SOE investments on an ongoing basis, actively explore

PPPs, link the target capital structure and dividend of each SOE to its business plan, and strengthen the coherence of SOE oversight by consolidating this responsibility with the minister of finance.

Progress on the implementation of the policy has been uneven, however, and only some its principles have been reflected in draft amendments to the SOE Act and regulations endorsed by Cabinet in 2022. In 2018/19, the government explored the merits of establishing a PPP Unit and project screening process, and identified potential PPP project opportunities, but none of these were pursued because of factors including public financing constraints, development partner program design, and a lack of commitment from counterpart public agencies. The PPP Unit, which was to be established in the Ministry of Finance, did not progress because of lack of resources, but the ministry has been actively involved in the development of the country's first large PPP project, the Tina River hydroelectric plant, which reached financial close in 2019. In this way, expertise in PPP project structuring and fiscal risk management is being established. Other PPP opportunities, such as in the ports, power, and water sectors, should be explored systematically in keeping with the policy.

The establishment of the Solomon Islands Airports Corporation under the Companies Act and SOE Act is a positive step towards commercialization, and could facilitate future partnerships with the private sector. The SOE's commercial structure will also introduce greater transparency and accountability, enabling the Ministry of Finance to exercise its ownership monitoring function.

Ongoing review of the rationale and VFM of the state's investment in SOEs, as prescribed by the policy, will require sustained political commitment. It is not clear whether recent decisions to recapitalize the Development Bank of Solomon Islands, which had been under liquidation, and further invest in the Commodities Export Marketing Authority (CEMA), the primary activity of which is regulatory, were the outcome of an assessment of alterative modalities for achieving the target outcomes. Similarly, a cost/benefit analysis of maintaining Investment Corporation Solomon Islands, which is a holding company for SOE shares and other government assets, would be warranted, particularly since it plays no monitoring or governance role over the SOEs, yet carries the costs of running a board and executive staff.

6. SIEA and Climate Change

SIEA (trading as SolPower) is a vertically integrated state-owned power utility which produces all of the grid-connected power consumed in Solomon Islands. In addition to the capital city of Honiara, SolPower supplies eight provincial centers on separate islands. Access rates are low, however, with only 9% of the country's population connected to the grid. The access rate is 64% in urban Honiara, and 3% in rural areas. About one third of all households have small solar home systems.

SolPower's reliance on diesel generation has not changed in recent years, but is set to fall from 100% to 10% by 2030. In 2020, 98% of power generated was from diesel fuel, and 2% from solar. SolPower is working to implement the government's climate change mitigation and adaptation policies, which require 100% of electricity produced and supplied to be renewable for Honiara by 2030, and for the rest of the country by 2050.[97] SolPower is on track to source 58.5% of the power supplied to the country from hydro, 31.6% from solar, and 9.9% from diesel by 2030.

SolPower's renewable energy targets are driven primarily by cost considerations, a desire to reduce dependency on imported fuel, and compliance with the government's climate change commitments. SolPower has the highest electricity tariffs in the Pacific because of its reliance on imported diesel fuel and absence of tariff subsidies.[98] The construction of the 15MW Tina River hydroelectric plant, which will supply an estimated 68% of Honiara's consumption, is expected to reduce tariffs by 35%, from $0.51 per kilowatt-hour in 2019 to $0.33 per kilowatt-hour when the plant is commissioned.[99] In addition to the Tina River hydropower plant, SolPower is building several hybrid solar/diesel plants in the provincial centers, 10 of which are being financed with support from development partners.

A core challenge to SolPower's transition to renewable energy has been access to land and expertise. The Tina River Hydropower Project has taken more than a decade to develop, as agreements with landowners were negotiated and concessional financing secured, and similar land challenges exist for solar farms. The Tina River Hydropower Project has been structured as a PPP, under a 34-year power purchase agreement, with a concurrent land lease with the various landowners. In the provinces, the development of hybrid solar/diesel plants has also been delayed by the difficulties of securing land.

97   Government of Solomon Islands. 2016. *National Development Strategy, 2016–2035*. Honiara.

98   SolPower's tariff is based on full cost recovery, and is adjusted as costs change. A new tariff regulation came into effect in August 2021.

99   https://www.adb.org/sites/default/files/project-documents/50240/50240-001-rrp-en.pdf.

Figure 40: SolPower Profitability 2010–2021

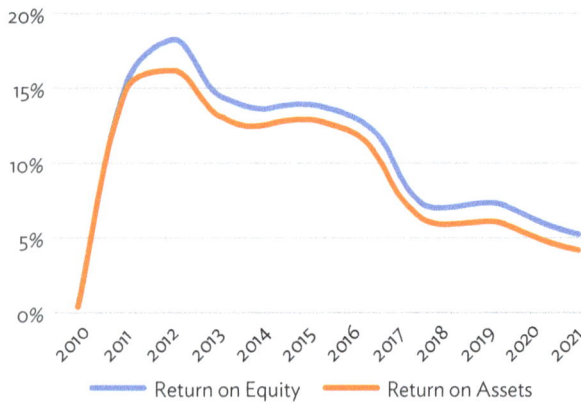

Source: Solomon Islands Electricity Authority audited accounts

SolPower is one of the most profitable power utilities in the Pacific, averaging an ROE of 10% and ROA of 9% over the 2010–2021 period. Profitability decreased after a tariff recalibration in 2017, and an agreement with the government to discontinue CSO payments. The company has accumulated healthy cash reserves and, as of 2021, held only S$18 million in interest-bearing debt. This places it in a strong position to withstand future shocks, including climate change-induced damage to infrastructure. In 2020, as part of the government's response to COVID-19, SolPower reduced its tariff and suspended disconnections, which caused its ROE to drop to 6% and ROA to 5%. This trend persisted in 2021 with ROE of 5% and ROA of 4%. SolPower paid dividends to the state of S$5 million in 2020 and S$4 million in 2021, and holds S$70 million in government bonds.

SolPower recognizes the physical risks to its assets posed by climate change, and is developing a resilience and adaptation strategy. While future investments in transmission and generation infrastructure are expected to incorporate climate-resilient designs, existing infrastructure remains vulnerable. Like PNG, Solomon Islands is prone to natural hazards such as earthquakes, volcanic eruptions, tsunamis, landslides, and storm surges. Currently, SolPower maintains the spares and equipment needed to repair recurrent storm-related damage to transmission lines, but a broader strategy will enable it to incorporate climate change risks into its investment program, and disclose associated business risks in its annual reports.

H. Tonga

1. SOE Portfolio Composition and Contribution to GDP

Tonga has 12 active SOEs involved in a range of commercial activities including utilities, transport, and communications. Tonga Power, Tonga Communications Corporation (TCC), Tonga Airports, and Tonga Cable represented 78% of the portfolio's total assets in 2020. The portfolio has remained unchanged since 2015, when the Tonga Development Bank was placed under the oversight of the Ministry of Finance and Tonga Forest Products was privatized.

In 2020, these 12 SOEs represented between 23% and 30% of the country's total capital stock, yet contributed 6% to GDP. The contribution of Tonga's SOEs to GDP has remained relatively stable over the 2010–2020 period, averaging 6%.

Figure 41: State-Owned Enterprise Portfolio Assets 2020 (T$531 million)

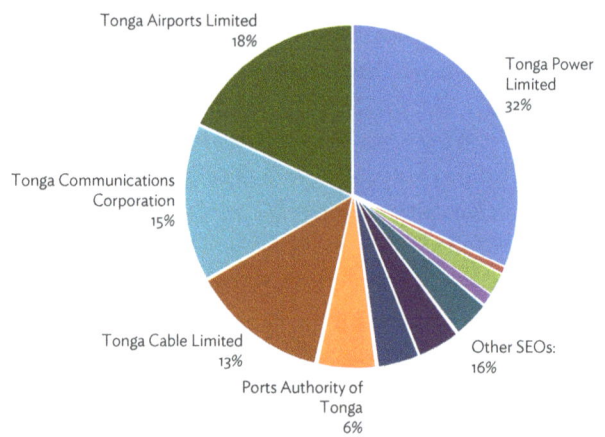

Source: State-Owned Enterprise audited accounts

A new SOE, Lulutai Airlines, was registered in May 2020 and is 100% owned by the state, but is not regulated by the PE Act. Its five-member board, composed entirely of Cabinet ministers, could only serve as directors for the first year of operation if it was regulated by the PE Act. Given the complexity and risks associated with operating an airline, it would be prudent to ensure that Lulutai Airlines has experienced management and a professional board as soon as possible, and operates under a clear commercial mandate. Placing it under the PE Act and the supervision of the Ministry of Public Enterprises would support this outcome.

2. SOE Performance 2010–2020

The portfolio's average returns have dropped sharply since 2018 as escalating costs have outpaced revenue growth, exacerbated by COVID-19. Average ROE was 4.7% and ROA was 2.9% during the 2010–2018 period, and dropped to 3.1% and 1.6% respectively in 2019–2020.

Figure 42: State-Owned Enterprise Portfolio Profitability 2010–2020

Source: State-Owned Enterprise audited accounts

While some SOEs suffered a contraction in revenues because of COVID-19 restrictions, such as Tonga Airports Ltd, Friendly Islands Shipping Agency, and Ports Authority Tonga, the sharp decline in profitability at Tonga Power Ltd (TPL) and TCC were more surprising, as demand for their outputs remained strong. In the case of TPL, the absence of tariff compensation from the government in 2020, together with increased maintenance and depreciation costs on new assets, resulted in the company's first loss-making year since 2009. TCC, which had an ROE of 7% and contributed 31% of the portfolio's net profit in 2019, suffered a 6% drop in revenue in 2020 while costs Tonga increased 4%, yielding an ROE of 0.3%. This contrasts

with Cable Ltd, which saw its profitability sharply increase in 2020, contributing 48% of the SOE portfolio's total profit.

Figure 43: Net Profit 2015–2020 (T$'000s)

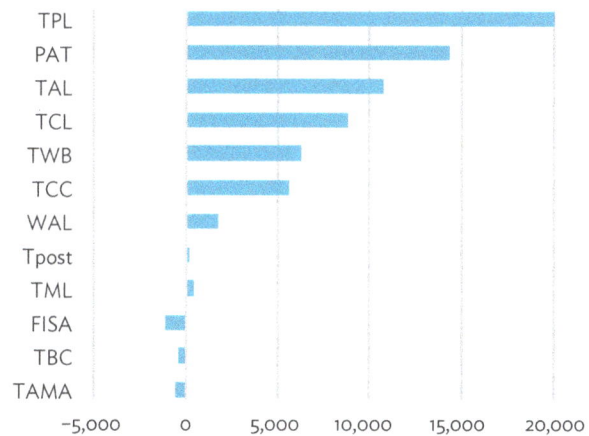

TPL = Tonga Power Limited, PAT = Ports Authority of Tonga, TAL = Tonga Airports Limited, TCL = Tonga Cable Limited, TWB = Tonga Water Board, TCC = Tonga Communications Corporation, WAL = Waste Authority Limited, Tpost = Tonga Post Limited, TML = Tongatapu Markets Limited, FISA = Friendly Island Shipping Agency, TBC = Tonga Broadcasting Corporation, TAMA = Tonga Assets Managers and Associates Ltd.

Source: State-Owned Enterprise audited accounts

More than 50% of the portfolio's net profits from 2015 to 2020 were provided by TPL and PAT. As with many of the SOE portfolios in this benchmarking study, Tonga's smaller SOEs continue to struggle with marginal profitability. Asset utilization remains low, averaging 34% over the 2010–2020 period, and the cash/current liabilities ratio sharply declined after 2015, averaging 0.74 from 2015 to 2020 compared to 2.13 from 2010 to 2014. This sharp decline reduced the liquidity of the portfolio, increasing its vulnerability.

3. Preliminary 2021 Results

SOE portfolio financial performance improved markedly in 2021 as compared to 2020, with net profits up 85%.[100] Consolidated ROE increased from 2.2% to 4.1% and consolidated ROA increased from 1.1% to 2.1%. While portfolio revenues declined by 2%, they were more than offset by a 7% reduction in costs. TPL recorded a net profit, with ROE rising from −1% in 2020 to 2% in 2021. Despite the impacts of COVID-19, energy demand increased by 3.3% during the reporting period, while generation and maintenance costs declined. Other notable improvements include Tonga Communications Corporation, with ROE rising from 0% to 2%, and Tonga Water Board, whose 2%, decline in revenue was offset by a 12% reduction in costs, resulting in a 35% increase in profitability.

---

[100] This data is for the 10 state-owned enterprises for which 2021 audited accounts were available as of November 2022. The two that are excluded are Friendly Islands Shipping Agency and Tonga Airports Limited.

4. COVID-19 Impact and Response

While most SOEs managed to weather the first 4 months of the COVID-19-induced economic contraction without generating a loss, all but four (Tonga Water Board, Waste Authority Ltd, Tonga Post Ltd, and Tonga Cable) saw a decrease in revenue in fiscal year 2020. This trend improved in 2021, with only four SOEs experiencing a decline in revenues compared to 2020, and at least two of these managing to remain profitable. In 2020, the government provided financial support to the more fragile SOEs involved in providing essential services, including T$1 million in grants to Friendly Islands Shipping Agency, Tonga Airports, Tonga Broadcasting Commission, and Tonga Markets Corporations Ltd. CSO funding was also provided to Friendly Islands Shipping Agency, Tonga Broadcasting Commission, and Waste Authority. TPL, which received T$3.2 million in 2019 to offset lost revenue from a tariff cap, received no further funding from the government in 2020 or 2021. Beyond the initial financial impact on SOEs, COVID-19 also delayed the beginning of new projects and supply of essential spare parts, likely lengthening its broader impact.

5. Legal, Governance, and Monitoring Framework

Tonga has one of the strongest SOE governance, monitoring, and disclosure frameworks in the benchmarking sample. The Ministry of Public Enterprises (MPE) was established in 2007, with the minister of public enterprises as the sole responsible minister. The SOE Act, in place since 2002, was amended in 2007, 2010, and 2020, each time to strengthen governance and accountability provisions. SOEs are required to prepare corporate plans, which include CSO costs, and submit audited accounts within 6 months of the end of the financial year. Ten out of 12 SOEs complied with these requirements in 2020. The MPE has also implemented a standardized reporting format for SOEs, and developed SOE-specific profit targets based on a simplified capital asset pricing model. Tonga remains the only PacDMC that publishes summaries of SOE audited annual accounts in local newspapers, reporting performance against targets in the corporate plan. SOE audited accounts are also available on the MPE's website, along with consolidated portfolio results.

The Ministry of Public Enterprises uses shared SOE boards and a codified, skills-based process for appointing SOE directors. SOEs with similar attributes such as the Utilities group (TPL, Tonga Water Board, and Waste Authority), share a common board, situate themselves in the same headquarter building, and use shared services such as billing and collections. In 2020, Cabinet endorsed new SOE director selection guidelines, incorporating a skills-based approach run by the MPE and a Cabinet subcommittee, and requiring the minister to appoint candidates shortlisted through this process. In 2021, this process was incorporated into regulations under the PE Act, which are currently pending. These are positive steps to ensure that SOE boards have the skills and experience needed to exercise their stewardship role.

6. SOE Reform Highlights 2015–2022

The establishment of three PPP contracts since 2016 has provided valuable PPP structuring experience to the MPE, and sent a signal to investors that Tonga is ready for PPPs and open to foreign investors. While the first PPP contract, for a 2MW solar farm, was not competitively tendered, the second, larger 6MW plant was the subject of an international tender that yielded several competitive proposals, resulting in a contract with a New Zealand-based operator. Further PPP opportunities exist in the power, water, waste, and transport sectors, including through the contracting out of CSOs. These opportunities should be pursued as an integral part of the ongoing implementation of the PE Act.

Despite Tonga's robust governance and monitoring framework, maintaining commercial returns requires ongoing political commitment at the highest level. Pressure continues to be placed on some SOEs, most notably TPL, to provide services below their true cost. In 2018, Cabinet endorsed the structuring of a concession for the cargo handling operations at the Queen Salote International Wharf, but suspended the competitive tender before it could be completed. A second attempt at the tender in 2020/21 also incurred delays, but a contract was ultimately awarded in 2022. Cabinet also created an SOE airline in 2020 without placing it under the oversight of the MPE.

Box 12: Tonga Reform Highlights 2015–2022

Public–private partnerships: Two solar independent power producer contracts with Tonga Power; one concession for Tonga Forest Products Ltd assets and plantation on 'Eua Island.

Legislation: Amended State-Owned Enterprise Act (2020) and regulations under the Act (2021).

Governance: State-owned enterprise director selection guidelines endorsed by Cabinet.

Source: Pacific Private Sector Development Initiative.

7. Tonga Power and Climate Change

TPL is a vertically integrated state-owned power utility which produces and delivers about 90% of the power consumed in Tonga. It is subject to the government's climate change mitigation and adaptation policies, most notably the renewable energy targets articulated in the Tonga Energy Roadmap (TERM) and Tonga Strategic Development Framework II. TPL has incorporated the TERM target of 50% diesel fuel savings by 2020 and Strategic Development Framework target of 50% renewable energy by 2025 into its business plan, and is working towards TERM's goals of 70% renewable electricity by 2030 and 100% by 2035. While TPL has not yet achieved the 2020 goal of 50% diesel fuel savings,[101] it has increased the share of renewable sources of fuel from 8% in 2015 to 13% in 2020. In 2021, renewables accounted for 11.5% of total generation.

Figure 44: Tonga Power Limited Fuel Sources 2015 and 2020

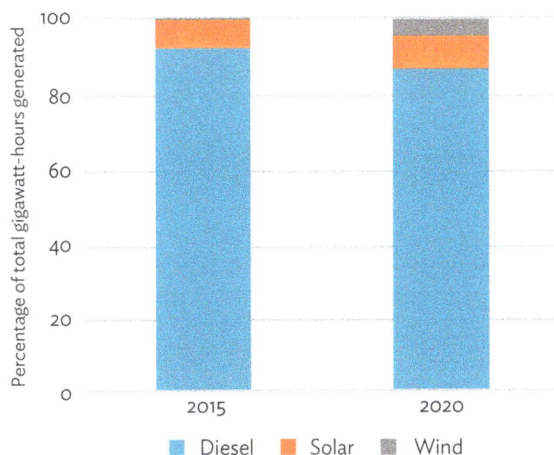

Source: Tonga Power Limited

TPL's renewable energy targets are primarily driven by cost and supply considerations, but are also influenced by a desire to reduce carbon emissions. TPL's feasibility studies indicate its renewable energy targets of 50% and 70% would—over the lifetime of the new asset life cycles—contribute 1.49 terawatt-hours of renewable energy, save 301 megaliters of diesel fuel, and emit 808 kilotons less carbon dioxide.

Challenges to achieving the renewable energy targets include the upgrading of network infrastructure, installation of battery storage capability, timely completion of land negotiations, and impact of tropical cyclones. While TPL is upgrading its infrastructure and battery capacity, these projects have incurred COVID-19-related delays in mobilizing equipment and expertise. Tropical cyclones have intensified in recent years, with Cyclone Gita causing an estimated T$17.41 million in damage to TPL's assets in 2020. TPL is adapting to extreme weather events by investing in its network infrastructure. For example, utility power poles are being strengthened, stronger and more resilient ABC bundle cables added, and underground service lines installed.

Figure 45: Tonga Power Limited Profitability 2010–2021

Source: Tonga Power Limited audited accounts

TPL has experienced substantial fluctuations in profitability over the past decade, linked in part to fluctuations in diesel import prices. While the transition to solar and wind is reducing generation costs, TPL cannot maximize savings until the battery storage capacity is in place. TPL is also investing in grid upgrades to improve distribution. Together with the cost of repairing damage to power infrastructure caused by Cyclones Gita in 2018 and Harold in 2020, this drove up overall costs in recent years.

[101] For fiscal year 2020, Tonga Power Limited estimated the fuel saving from renewable energy at 2.3 million liters, for a total fuel displacement of 14.5%.

In 2020, the government did not compensate TPL for the tariff cap, as it had in 2018 and 2019. Instead, it agreed to a lower return on investment. This approach carried over into 2021, when an agreed tariff increase was not passed on to consumers, with the government recognizing the tariff subsidy as a CSO and agreeing to a lower return on investment from TPL.

TPL will continue to displace diesel generation with renewable energy, and will do so through partnerships with the private sector. Some 90% of the growth in solar power generation since 2015 was through IPP contracts, and TPL expects to add another 24MW of solar by 2025. Since 2015, TPL has developed expertise in the development and implementation of competitive IPP tenders and negotiation of power purchase agreements. Competitive tenders have been run for (i) two battery energy storage systems awarded in 2019, (ii) a 6MW solar IPP that was commissioned in September 2022, and (iii) a 24MW solar IPP with 24MW/41MWh battery which was tendered in 2022 in two lots and will take Tonga to 70% renewable energy penetration. These openly competitive tenders ensure that TPL obtains the best value for money for its infrastructure investments.

In a departure from this approach, a second 6MW solar PPA contract was signed in November 2019 without a tender process. The PPA has not yet reached financial close as of November 2022 because a number of conditions remain outstanding. Given the responsibility of TPL directors to act in the best commercial interest of TPL, it will be incumbent on them to demonstrate this unsolicited proposal presented unique features that made a competitive tender impossible. They must also outline how the resulting terms of the PPA present at least as much VFM as relevant benchmarks, such as the first 6MW solar IPP.

I. Vanuatu

1. SOE Portfolio Composition and Contribution to GDP

Vanuatu's SOE portfolio comprises seven active SOEs— Airports Vanuatu (AV), Air Vanuatu (Operations) Limited (AVOL), National Bank of Vanuatu (NBV), Vanuatu Agriculture Development Bank (VADB), Vanuatu Broadcasting and Television Corporation (VBTC), Vanuatu Post Limited (VPL), and National Housing

Corporation (NHC). Vanuatu Post has not published audited accounts since 2017, and NHC since 2014. The absence of financial data for VPL will impact portfolio performance analysis. Vanuatu Post averaged an ROE of 6% in the period 2010–2017 and generated 8% of total portfolio profits in 2017. The lack of financial data for NHC is unlikely to materially impact analysis of the portfolio's performance, as the SOE has not traded for several years.

Figure 46: State-Owned Enterprise Portfolio Assets 2020 (Vt33.6 billion)

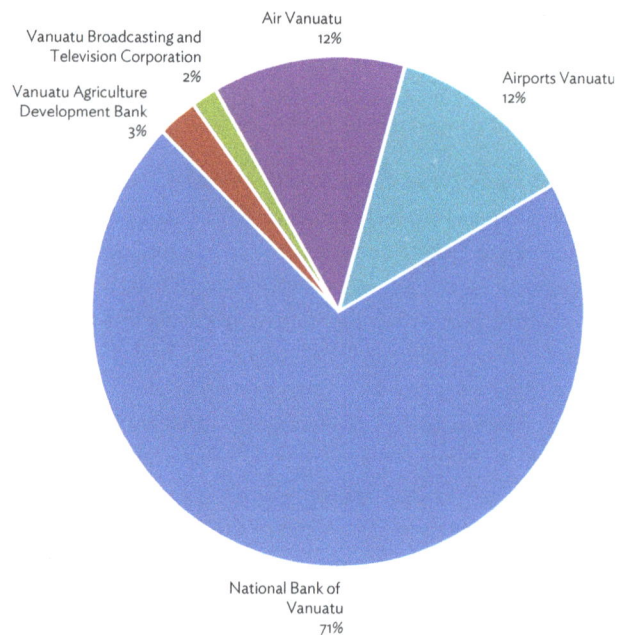

Source: State-Owned Enterprise audited accounts

Unlike other countries included in this benchmarking study, Vanuatu does not have SOEs providing power and water services. In the larger urban areas of the country, these services are provided by the private sector through long-term concession agreements. In other areas, the services are provided by the departments of energy and water resources.

The National Bank of Vanuatu represented 71% of total portfolio assets in 2020. The portfolio is not large in number or asset size, comprising 6% of the economy's total fixed assets, broadly in line with SOEs' contribution to

GDP, which has averaged 2.6% in the period 2010–2020. However, the SOE portfolio made a –0.1% contribution to GDP in 2020—effectively shrinking GDP—because of significant losses at Air Vanuatu (Vt2.3 billion) and Airports Vanuatu (Vt0.45 billion).

2. SOE Performance 2010–2020

In the period 2010–2020, the portfolio's average ROE was –11.5% and average ROA was –1.2%. The main cause of the poor performance has been Air Vanuatu, which accumulated losses of Vt5.4 billion in the decade to 2020. The government has provided significant financial support to AVOL, including a 2019 loan of more than Vt2 billion, which was converted into equity in that year. The government also provided a guarantee of Vt592 million and cash injection of Vt200 million in 2020. Despite the significant financial support, the airline's shareholders' funds by 2020 were –Vt3.2 billion and the debt ratio[102] was 88%, indicating the

Figure 47: State-Owned Enterprise Portfolio Profitability 2010–2020

Source: State-Owned Enterprise audited accounts

SOE is on the brink of financial failure. A commission of enquiry was established in 2020 and the government took direct control of AVOL in March 2021. The airline was in the process of purchasing four new aircrafts from Airbus, estimated at a cost of $185 million—about 20% of GDP. The SOE's board has changed three times between 2017 and March 2021; the board appointed in 2020 was replaced by a new board in 2021.

The portfolio return in the second half of the decade was weaker than the first. In the period 2015–2019, average

portfolio ROE was –9.6%, while ROA was –1.2%. Because of the combined losses of Air Vanuatu and Airports Vanuatu, these fell to –81% and –7% respectively in 2020.

Three SOEs have historically generated almost all profits: AV, NBV, and VADB. While the portfolio accumulated Vt3.6 billion in aggregate losses in the period 2015–2020, NBV contributed profit of Vt453 million and VADB Vt409 million from 2015 to 2020. Airport Vanuatu generated Vt519 million in profits in the period 2015–2019, but the 2020 loss brought the aggregate contribution for the period 2015–2020 to a modest Vt73 million.

Figure 48: Net Profit 2015–2020 (Vt million)

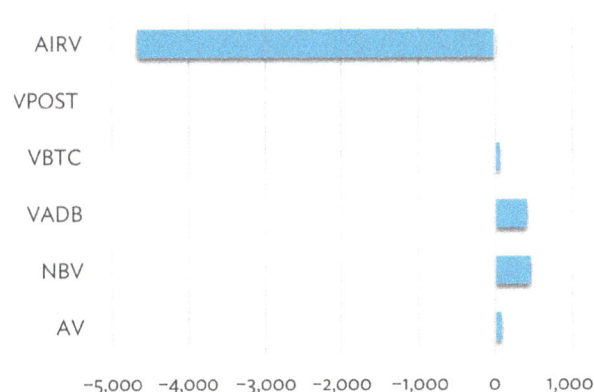

AIRV = Air Vanuatu, VPOST = Vanuatu Post, VBTC = Vanuatu Broadcasting and Television Corporation, VADB = Vanuatu Agriculture Development Bank, NBV = National Bank of Vanuatu, AV = Airports Vanuatu.
Source: State-Owned Enterprise audited accounts

3. COVID-19 Impact and Response

The government implemented a stimulus package in 2020 to support businesses by deferring several taxes and charges, such as road taxes, work permit fees, business license fees, resident permit charges, and rent taxes. While the combination of Cyclone Harold and COVID-19 had an adverse impact on the economy, with GDP contracting 8.5% and government revenue down 2.9% on the prior year, the government did not seek subsidies from SOEs to contribute to the stimulus.

AVOL and AV were the most impacted by COVID-19. Travel restrictions added to the financial strains impacting AVOL. In 2020, revenues at AVOL were down 68% and AV 54% compared to 2019. The significant and rapid drop in passenger movements in 2020 resulted in a Vt464 million negative movement in AV's net profit for the year.

---

[102] Total debt to total assets.

NBV's profitability declined in 2020 from an average ROE of 23% in 2018 and 2019 to 3% in 2020. The COVID-19-affected tourism sector is a major customer base for the bank, prompting a 13.5% decline in the bank's interest income in 2020 compared with 2019. Other expenses also increased 8.8% mainly because of an increase in doubtful loans.

Figure 49: Passenger Movements Port Vila, August 2019 to August 2021

Source: Airports Vanuatu website

Conversely, VADB's profitability increased between 2019 and 2020 as it improved the quality of its loan portfolio. Loans increased 15% and loan income 25%, while provisioning for doubtful loans was reduced by 205%, resulting in a 18% ROE, above the average ROE of 8% for the preceding 4 years. The difference in trading results for the two financial institutions could be attributed to their different client base, VADB's exposure to the domestic agriculture sector, and VADB's limited exposure to the tourism sector.

VADB maintains high board expenses compared to industry benchmarks. In 2020, VADB's board expenses totaled Vt5.5 million, more than 10% of total employee costs for that year. While this was down from Vt7.9 million in 2019, it remains high by industry benchmarks. For example, directors' fees for NBV in 2020 were Vt3.2 million or 0.6% of total employee costs, Development Bank of Samoa directors' fees were 4% of total employee costs, and directors' fees at DBK were 7% of total employee costs. VADB accounts do not provide a breakdown of what comprises board expenses. Board expenses that exceed 10% of total employee costs should be explained and justified.

4. Legal, Governance, and Monitoring Framework

Vanuatu has the weakest legal, governance, and monitoring framework of the countries participating in this study. While most of the SOEs are registered companies under the 2012 Companies Act, there is no framework for the effective exercise of the government's shareholder powers. A comprehensive SOE ownership policy was adopted by Cabinet in October 2013, but implementation has been patchy. The policy was designed to form the basis of an SOE law, and work started on drafting an SOE bill in 2014. The bill was submitted to Parliament in June 2018 but was referred to a parliamentary ad hoc committee for review. A number of changes were requested by the ad hoc committee, but a change of government resulted in a new ad hoc committee review in June 2021. By the end of 2021, the ad hoc committee's views had been incorporated into an updated draft awaiting the Minister of Finance's approval to submit the bill to Parliament, but this was not done before the October 2022 election. As of November 2022, the bill remains pending. The bill contains many good practice principles—it requires a measurable commercial mandate, provides a CSO framework, assigns one minister responsibility for all SOEs, strengthens and clarifies director duties, requires the preparation and publication of forward-looking business plans, and strengthens accountability through increased transparency. The bill would also enhance the powers of the SOE ownership monitor within the Ministry of Finance to gather information from SOEs and provide policy advice to the minister.

Successive reviews of the SOE Bill have led to a diminution of some of its good practice principles. For example, the second ad hoc committee review expanded the power to appoint public servants to SOE boards, and introduced the requirement that directors be paid based on meeting fees, rather than a set annual fee. This creates the incentive to hold numerous board meetings, even multiple meetings on the same day. The ongoing fine-tuning of the SOE bill by politicians is delaying its submission to Parliament, and introducing uncertainty in some provisions.

SOE ownership monitoring is virtually non-existent. A SOEMU has been established within the Ministry of Finance, the Government Business Monitoring and Evaluation Unit (GBMEU). However, there is no legal requirement for SOEs to provide the GBMEU with information, whether financial or operational. For much of the past decade, the GBMEU has comprised just one staff member. In Vanuatu, SOEs seldom produce forward-looking business plans: the GBMEU reports they have received one business plan across the entire SOE portfolio in 2018, and nothing since. Governance arrangements are weak. Line ministers appoint directors to the boards of SOEs operating within the ministry's economic sector. Directors often change as ministers change, and there are no director performance reviews undertaken. The government should move to enact the SOE bill and resource the SOEMU as a matter of urgency.

5. SOE Reform Highlights 2015–2022

Following adoption of the SOE Ownership Policy in 2013, a reform program was commenced and continued through to 2015. The program developed reform strategies for eight SOEs identified by the government as priorities. The reform strategies included identifying and costing CSOs for AVOL, assessing the merits of entering joint ventures with local landowners through Metenesel Estates Limited, liquidating three non-operational SOEs, and assisting in the commercialization of the balance of SOEs. While implementation by the government has been patchy—the three non-operating SOEs have not yet been liquidated—there has been some success, notably the increased commercial focus of VBTC. VBTC's ROE increased from an average of –2% in 2015–2017 to 7% in 2018–2020.

The Vanuatu National Provident Fund increased its shareholding in NBV from 15% to 56% in September 2020, while the government acquired the 15% owned by International Finance Corporation, bringing the government's share down from 70% to 44%. Removing NBV from the SOE portfolio will significantly reduce the portfolio's size in terms of assets controlled. The name of Vanuatu Agriculture Development Bank (VADB) was changed to Vanuatu Rural Development Bank in June 2021, with the intention to expand the services it provides to Vanuatu's rural community. However, an administrator was appointed by the government in October 2021 following the discovery of financial irregularities in VADB by an external auditor.

An investment vehicle, Interchange Limited, was formed in 2014 between the government, Vanuatu National Provident Fund, VPL, Fidelity, and Interchange Holdings

Limited which constructed Vanuatu's first submarine internet cable. The cable connected Port Vila to Suva, where it connects to the South Pacific cable. A second cable is planned to connect Vanuatu to Solomon Islands, and the joint venture is also building satellite connectivity. The joint venture investment appears to have generated real benefits, with wholesale prices declining by 95% and capacity usage increasing over 25 times since the cable's completion. The joint venture has also paid public shareholders— government, VPL, and Vanuatu National Provident Fund— more than Vt70 million in dividends in fiscal years 2020 and 2021.

In July 2021, the government released a tender for the operations of the Malekula (the second-largest island) and Tanna island power assets for 20 years. This would expand the role of the private sector in delivering power to four of the country's largest islands, representing 45% of the population. The Department of Energy remains responsible for delivering power to the rest of the population, noting that 62% of the total population has access to grid power, with many rural households generating their own solar energy.

Given the lack of an effective SOE governance framework, it is fortunate that the SOE portfolio is comparatively small in terms of GDP, limiting the adverse economic impact from its poor financial performance. The portfolio is heavily weighted to the aviation sector following the sale of NBV, exposing the government to risks it has proven incapable of managing. The government should look for opportunities to introduce greater private sector participation in the ownership and operation of the remaining SOEs, as it has successfully done in the power and water sectors.

6. Electricity Sector and Climate Change

Vanuatu has set an ambitious target of generating 100% of its electricity through renewable sources by 2030.[103] This will take considerable effort and investment given the current rate of renewable use, and may be impacted by the terms of the existing PPP contracts with the two major private utility companies, which manage government-owned assets. Vanuatu Utilities and Infrastructure Limited operates the Luganville electricity concession on Espiritu Santo (the largest island), and Union Electrique du Vanuatu operates a concession in the most populous island of Efate (99,800 people). If these contracts include incentives for increased renewable energy use, the transition will be facilitated.

---

[103]   Government of Vanuatu. 2016. *Vanuatu National Energy Road Map*. Port Vila.

Figure 50: Share of Renewable Energy in Electricity Production 2010–2020

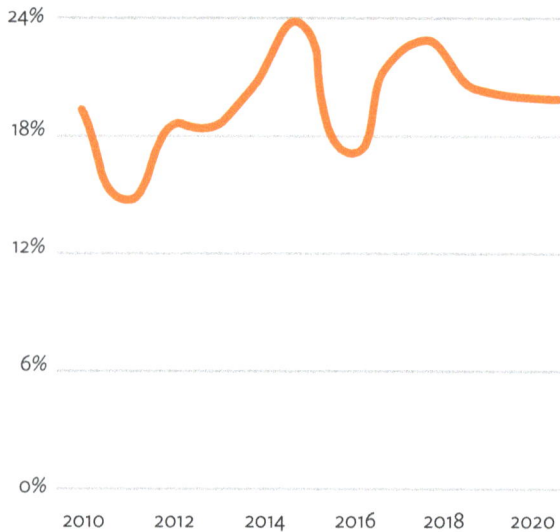

Source: Enerdata (www.enerdata.net)

Similarly, the future contract for the operations of the Malekula and Tanna island power assets could include incentives for increased renewable energy generation. Vanuatu's experience with PPPs in the energy, water, and ports sectors should provide it with an opportunity to structure terms that align with its renewable energy goals. The greater challenge may be to expand access to renewable energy to the remaining 38% of the population in a sustainable manner.

Appendix: State-Owned Enterprise Performance Indicators

Table A1: Fiji State-Owned Enterprise Performance Indicators, FY2020

State-Owned Enterprise		Government Ownership (%)	Return on Equity (ROE) (%)	Return on Assets (ROA) (%)	Total Assets (F$'000)	Total Revenue (F$'000)	Asset Utilization (%)	Total Liabilities (F$'000)	Average ROA FY2015–FY2020 (%)
FMIB	Fiji Meat Industry Board	100	(1)	(1)	34,823	3,121	9	1,490	2
PRB	Public Rental Board	100	6	4	42,124	4,949	12	15,441	2
HA	Housing Authority	100	3	1	206,819	17,567	8	147,952	1
UTOFML	Unit Trust of Fiji (Management) Limited	100	17	12	8,206	3,127	38	2,574	22
FPFL	Food Processors (Fiji) Limited	100	27	12	6,792	4,719	69	3,791	3
FRL	Fiji Rice Limited	100	–	40	6,256	4,958	79	7,671	10
YPCL	Yaqara Pastoral Company Limited	100	37	12	22,421	4,599	21	15,116	14
FPTCL	Fiji Public Trustee Corporation Limited	100	0	0	17,479	1,699	10	223	4
FP	Fiji Pine	100	26	23	230,408	126,349	55	28,064	4
PAFCO	Pacific Fishing Company Limited	100	18	8	45,498	43,376	95	24,156	1
Airpac	Air Pacific (Fiji Airways)	51	(88)	(10)	1,170,840	184,535	16	1,033,002	2
FSC	Fiji Sugar Corporation	68	–	10	152,256	153,271	101	356,232	(11)
EFL	Energy Fiji Limited	80	7	5	1,157,199	271,466	23	438,276	5
PFL	Post Fiji Limited	100	9	3	40,476	12,037	30	24,897	3
AFL	Airports Fiji Limited	100	0	0	568,181	54,699	10	124,222	12
FBCL	Fiji Broadcasting Corporation Limited	100	11	6	46,083	19,885	43	21,549	7
ATS	Air Terminal Services	51	(7)	(6)	22,974	7,062	31	2,133	5
FDB	Fiji Development Bank	100	1	0	540,331	51,712	10	366,997	1
CMFL	Copra Millers of Fiji	100	2	1	9,244	3,789	41	2,505	(2)

Source: SOE Annual Accounts

Table A2: Kiribati State-Owned Enterprise Performance Indicators, FY2020

State-Owned Enterprise		Government Ownership (%)	Return on Equity (ROE) (%)	Return on Assets (ROA) (%)	Total Assets (A$'000)	Total Revenue (A$'000)	Asset Utilization (%)	Total Liabilities (A$'000)	Average ROA FY2015–FY2020 (%)
AK	Air Kiribati	100	7	3	12,105	13,751	114	6,984	(6)
AKA	Airports Kiribati Authority	100	(8)	(8)	18,609	1,699	9	4	(0)
BNL	Bwebweriki Net Limited	100	100	16	237	89	38	200	3
BPA	Broadcasting and Publications Authority	100	1	1	1,344	1,448	108	234	(4)
CPP	Central Pacific Producers	-	-	-	-	-	-	-	-
DBK	Development Bank of Kiribati	100	10	6	21,081	2,983	14	6,821	5
KHC	Kiribati Housing Corporation	100	3	3	12,613	3,542	28	346	0
KIC	Kiribati Insurance Corporation	100	4	4	9,600	2,430	25	575	5
KOC	Kiribati Oil Company	100	12	8	43,031	40,754	95	15,324	7
KPA	Kiribati Ports Authority	100	3	3	62,799	9,598	15	528	3
KSS	Kiribati Shipping Services	-	-	-	-	-	-	-	-
PVU	Plant and Vehicle Unit	100	30	30	13,057	6,818	52	(3)	9
PUB	Public Utilities Board	100	(1)	(1)	39,495	16,853	43	2,483	1
KCDL	Kiribati Coconut Development Ltd	100	9	9	6,001	5,895	98	91	8
KNSL	Kiribati National Shipping Line Ltd	100	(10)	(9)	4,024	2,586	64	478	(8)
TACL	Te Atinimarawa Company Ltd	100	3	1	5,063	2,274	45	3,549	1

Source: SOE Financial Accounts

Table A3: Marshall Islands State-Owned Enterprise Performance Indicators, FY2020

State-Owned Enterprise	Government Ownership (%)	Return on Equity (ROE) (%)	Return on Assets (ROA) (%)	Total Assets ($'000)	Total Revenue ($'000)	Asset Utilization (%)	Total Liabilities ($'000)	Average ROA FY2015–FY2020 (%)
KAJUR Kwajalein Atoll Joint Utilities Resources	100	–	–	–	–	–	–	(1)
MEC Marshalls Energy Company	100	31	17	34,833	35,793	103	15,713	14
NTA National Telecommunication Authority	100	15	4	26,551	10,269	39	18,882	2
RMIPA Marshall Islands Ports Authority	100	(6)	(6)	60,362	3,647	6	1,422	(1)
MIDB Marshall Islands Development Bank	100	11	10	46,012	7,269	16	3,035	9
MWSC Majuro Water and Sewer Company	100	–	(1)	2,870	1,751	61	3,520	(9)
MRI Majuro Resort	100	20	5	1,913	3,859	202	1,397	(2)
TCPA Tobolar Copra Processing Authority	100	37	23	3,435	2,508	73	1,328	6
AMI Air Marshall Islands	100	11	6	12,419	6,298	51	5,003	10
MAWC Majuro Atoll Waste Company	100	4	3	734	1,641	224	113	(15)
MISC Marshall Islands Shipping Corporation	100	(1)	(1)	9,255	4,401	48	1,003	5

Source: SOE Financial Accounts

Table A4: Palau State-Owned Enterprise Performance Indicators, FY2020

State-Owned Enterprise		Government Ownership (%)	Return on Equity (ROE) (%)	Return on Assets (ROA) (%)	Total Assets ($'000)	Total Revenue ($'000)	Asset Utilization (%)	Total Liabilities ($'000)	Average ROA FY2015–FY2020 (%)
BSCC	Belau Submarine Cable Corporation	100	31	2	23,573	2,997	13	21,789	2
PPUC	Palau Public Utilities Corporation	100	(7)	(3)	106,006	27,273	26	60,935	(3)
PNCC	Palau National Communications Corporation	100	(1,419)	(5)	31,134	12,796	41	31,027	6
NDBP	Palau National Development Bank	100	(1)	(1)	37,506	1,447	4	11,724	2

Source: SOE Financial Accounts

Table A5: PNG State-Owned Enterprise Performance Indicators, FY2020

State-Owned Enterprise		Government Ownership (%)	Return on Equity (ROE) (%)	Return on Assets (ROA) (%)	Total Assets ($K'000)	Total Revenue ($K'000)	Asset Utilization (%)	Total Liabilities ($K'000)	Average ROA FY2015-FY2020 (%)
TPNG	Telikom PNG Limited	100	0	0	1,246,340	389,909	31	814,225	(3)
PPL	PNG Power Limited	100	(5)	(2)	2,870,302	952,209	33	1,949,295	(1)
PPCL	PNG Ports Corporation	100	6	4	2,095,836	304,107	15	600,935	7
Post	Post PNG Limited	100	(1)	(1)	210,182	38,254	18	34,486	0
WPNG	PNG Water Board	100	4	2	583,368	118,002	20	250,704	3
ANL	Air Niugini Limited	100	(86)	(12)	814,198	620,310	76	701,618	(11)
MVIL	Motor Vehicle Insurance Limited	100	13	10	850,818	172,563	20	185,421	10
ERL	Eda Ranu Limited	100	17	13	256,358	119,344	47	54,592	(1)
NDB	National Development Bank	100	(3)	(2)	763,302	43,225	6	169,370	(1)
BMOB	BeMobile	100	–	(24)	213,961	163,372	76	406,826	(19)
KAL	Kumul Agriculture Ltd	100	(12)	(11)	46,045	592	1	3,196	(4)

Source: SOE Financial Accounts

Table A6: Samoa State-Owned Enterprise Performance Indicators, FY2020

State-Owned Enterprise		Government Ownership (%)	Return on Equity (ROE) (%)	Return on Assets (ROA) (%)	Total Assets (ST$'000)	Total Revenue (ST$'000)	Asset Utilization (%)	Total Liabilities (ST$'000)	Average ROA FY2015–FY2020 (%)
EPC	Electric Power Corporation	100	2	1	550,937	136,204	25	217,934	2
SWA	Samoa Water Authority	100	0	0	197,438	29,006	15	131,470	0
SAA	Samoa Airport Authority	100	2	1	178,614	23,163	13	121,316	0
SPA	Samoa Ports Authority	100	5	3	280,400	21,065	8	131,924	1
SSC	Samoa Shipping Corporation Limited	100	1	0	87,617	20,594	24	45,090	2
DBS	Development Bank of Samoa	100	(1)	0	170,140	9,452	6	111,064	0
SHC	Samoa Housing Corporation	100	6	3	62,359	8,203	13	27,232	3
PTO	Public Trust Office	100	8	4	10,961	1,320	12	5,782	2
SLC	Samoa Land Corporation	100	11	4	116,233	13,193	11	70,393	(2)
STEC	Samoa Trust Estates Corporation	100	(5)	(4)	47,346	507	1	2,038	(3)
SSS	Samoa Shipping Services Limited	100	(49)	(23)	2,422	2,716	112	1,284	(2)
PAL	Polynesian Airlines Limited	100	–	(33)	53,644	82,172	153	71,501	(18)
SAMPOS	Samoa Post	100	0	0	7,225	2,311	32	3,520	4
UTOS	Unit Trust of Samoa	100	18	13	3,767	1,977	52	1,125	21

Source: SOE Annual Accounts

Table A7: Solomon Islands State-Owned Enterprise Performance Indicators, FY2020

State-Owned Enterprise		Government Ownership (%)	Return on Equity (ROE) (%)	Return on Assets (ROA) (%)	Total Assets (SI$'000)	Total Revenue (SI$'000)	Asset Utilization (%)	Total Liabilities (SI$'000)	Average ROA FY2015–FY2020 (%)
SAL	Solomon Airlines Limited	100	–	(15)	203,352	195,445	96	250,811	(8)
SIAC	Solomon Islands Airport Corporation	–	–	–	–	–	–	–	–
SIPC	Solomon Islands Postal Corporation	100	–	–	–	–	–	–	–
SIBC	Solomon Islands Broadcasting Authority	100	3	3	31,177	10,510	34	730	4
SIEA	Solomon Islands Electricity Authority	100	6	5	1,514,116	452,552	30	223,098	8
SIPA	Solomon Islands Ports Authority	100	9	8	962,471	240,584	25	192,431	10
SIWA	Solomon Islands Water Authority	100	8	5	379,542	132,418	35	153,287	3

Source: SOE Annual Accounts

Table A8: Tonga State-Owned Enterprise Performance Indicators, FY2020

State-Owned Enterprise	Government Ownership (%)	Return on Equity (ROE) (%)	Return on Assets (ROA) (%)	Total Assets (T$'000)	Total Revenue (T$'000)	Asset Utilization (%)	Total Liabilities (T$'000)	Average ROA FY2015–FY2020 (%)
TPL Tonga Power Limited	100	(1)	0	167,752	67,469	40	105,797	2
TML Tongatapu Markets Limited	100	2	1	3,338	1,620	49	1,547	3
TIL Tonga Investments Limited	100	–	–	–	–	–	–	–
TMPL Tonga Machinery Pool Limited	100	–	–	–	–	–	–	–
FISA Friendly Island Shipping Agency	100	–	7	11,690	7,806	67	12,729	(3)
TBC Tonga Broadcasting Commission	100	(31)	(10)	6,752	1,777	26	4,666	(3)
TCC Tonga Communications Corporation	100	0	0	79,298	31,352	40	30,697	1
PAT Ports Authority Tonga	100	8	7	30,607	12,105	40	6,033	10
TWB Tonga Water Board	100	17	4	21,878	10,211	47	16,602	7
WAL Waste Authority Limited	100	29	10	6,573	4,884	74	4,399	7
TAL Tonga Airports Limited	100	(2)	(1)	96,150	11,729	12	59,997	3
Tpost Tonga Post Limited	100	11	2	19,314	1,496	8	15,571	1
TAMA Tonga Assets Managers and Associates Ltd	100	(3)	(3)	17,437	889	5	1,823	(1)
TCL Tonga Cable Limited	100	5	3	70,833	8,738	12	18,326	3

Source: SOE Annual Accounts

Table A9: Vanuatu State-Owned Enterprise Performance Indicators, FY2020

State-Owned Enterprise	Government Ownersh p (%)	Return on Equity (ROE) (%)	Return on Assets (ROA) (%)	Total Assets (Vt$'000)	Total Revenue (Vt$'000)	Asset Utilization (%)	Total Liabilities (Vt$'000)	Average ROA FY2015-FY2020 (%)
AV	Airports Vanuatu	100	(13)	(11)	4,164,095	429,496	10	694,169	0
AIRV	Air Vanuatu	100	-	(53)	4,238,646	2,420,937	57	7,413,871	(26)
NBV	National Bank Vanuatu	44	3	0	23,763,089	1,528,373	6	22,065,842	0
VADB	Vanuatu Agriculture Development Bank	100	18	17	877,182	203,477	23	49,589	10
VBTC	Vanuatu Broadcasting & Television Corporation	100	17	7	574,301	183,576	32	351,976	2
VPOST	Vanuatu Post	100	-	-	-	-	-	-	-
NHC	National Housing Corporation	100	-	-	-	-	-	-	-

Source: SOE Annual Accounts

www.ingramcontent.com/pod-product-compliance
Lightning Source LLC
Chambersburg PA
CBHW050048220326
41599CB00045B/7328